Rabbit Behaviour, Health and Care

Rabbit Behaviour, Health and Care

Marit Emilie Buseth

and

Richard A. Saunders
BSc (Hons) BVSc MRCVS CBiol MSB CertZooMed DZooMed (Mammalian) RCVS Recognised Specialist in Zoo and Wildlife Medicine
Rabbit Welfare Association & Fund Veterinary Adviser

www.cabi.org

CABI is a trading name of CAB International

CABI
Nosworthy Way
Wallingford
Oxfordshire OX10 8DE
UK

CABI
38 Chauncy Street
Suite 1002
Boston, MA 02111
USA

Tel: +44 (0)1491 832111
Fax: +44 (0)1491 833508
E-mail: info@cabi.org
Website: www.cabi.org

Tel: +1 800 552 3083 (toll free)
E-mail: cabi-nao@cabi.org

© M.E. Buseth and R. Saunders 2015. All rights reserved. No part of this publication may be reproduced in any form or by any means, electronically, mechanically, by photocopying, recording or otherwise, without the prior permission of the copyright owners.

A catalogue record for this book is available from the British Library, London, UK.

Library of Congress Cataloging-in-Publication Data

Buseth, Marit Emilie.
  [Store kaninboka. English]
  Rabbit behaviour, health and care / Marit Emilie Buseth, Richard A. Saunders.
    pages cm
  Includes bibliographical references and index.
  ISBN 978-1-78064-190-4 (alk. paper)
  1. Rabbits. I. Saunders, Richard A. II. Title.

SF453.B4713 2014
599.32--dc23

2013040387

ISBN-13: 978 1 78064 190 4

Commissioning editor: Sarah Hulbert/Rachel Cutts
Editorial assistant: Alexandra Lainsbury
Production editor: Simon Hill/Lauren Povey

Typeset by SPi, Pondicherry, India.
Printed and bound in Great Britain by Bell & Bain Ltd, Glasgow

In memory of Petter

Photo courtesy of Andy Purivance, USA

Contents

Foreword to the Norwegian Edition   ix
Glen Cousquer

Foreword   xi
Richard A. Saunders

Introduction   xiii

Acknowledgements   xv

1   The Origins and Development of Rabbits   1
2   The Rabbit as a Companion Animal   15
3   Behaviour, Learning and Communication   29
4   Social Rabbits   59
5   From Snout to Tail   83
6   Rabbit Nutrition   117
7   Neutering   139
8   Cleanliness and Hygiene   151
9   Rabbit Housing and Conditions   161
10   House Rabbits and Rabbit-proofing of the Home   177
11   Life Outdoors   187
12   Reproduction and Breeding Control   201

Epilogue   209

Bibliography   213

Index   219

Vilkanin. Photo courtesy of Hernan Vargas, USA

Foreword to the Norwegian Edition

As man developed from a hunter-gatherer and became increasingly dependent on pastoralism and agriculture, his relationship with animals changed. Dogs, horses, cattle, fowl, cats and rabbits were domesticated, some as food-producing animals, others as working animals. Over time, a number of these animals entered the home and were redefined as companion animals. Whilst the dog became known as man's best friend, whilst the writings of Anna Sewell cemented the place of the horse in our affections and the cat waltzes in and out of our lives as it chooses, the rabbit has crept up along the rails unnoticed. In today's modern society, people have less and less time for a dog or a horse and increasingly look to the rabbit for companionship.

The rabbit is not, however, a small dog, a funny looking cat or even a miniature horse and cannot be treated as such. As a species, it has evolved over thousands of years to fill a specific ecological niche. The nutritional, behavioural and physiological needs of the rabbit very much reflect this evolutionary process and an understanding of these needs is essential if rabbits are to be kept as companion animals.

Sadly, this understanding is all too often missing or deficient amongst rabbit owners. Many, perhaps even the majority, of the medical problems seen amongst domestic rabbits are directly attributable to poor husbandry and owner ignorance. Myiasis, dental problems and gastrointestinal disorders such as gut stasis all cause immense suffering amongst pet rabbits.

Owner education is essential if these welfare abuses are to become a thing of the past. Without a good understanding of how rabbits should be fed, housed, exercised and cared for, we cannot expect owners to meet their responsibilities as guardians of these delightful animals. In the UK, the Rabbit Welfare Association has committed itself to the education of the rabbit-owning public. It makes information on a wide range of husbandry and health issues freely and widely available. The Association has also funded residencies in rabbit medicine at a UK vet school, thus raising the profile of rabbit medicine within the veterinary profession. There still remains much work to do, however, for there are many owners out there who remain out of reach, who do not know where to look for information and advice and fail in their duty of care.

The work of a small number of rabbit enthusiasts can and does make a difference. This book bears testimony to this fact. Marit Emilie Buseth is one of Scandinavia's foremost advocates of rabbit welfare and has worked extremely hard to develop public awareness and understanding of rabbit health issues. Rabbits can make wonderful and rewarding companion animals. Caring for them and learning about them should not be an onerous task for it can greatly enhance the relationship between owners and their rabbits. This book will invest in this relationship, promoting an ethic of care that will not fail to bring its own rewards.

Glen Cousquer
BSc (Hons) BVM&S CertZooMed PGDOE MScMRCVS

Lago, a curious rabbit. Photo courtesy of Kaja Larsen Østerud, Norway

Foreword

I became aware of the first edition of this book when its author kindly gave me a signed copy when she attended a Rabbit Welfare Association & Fund conference 2 years ago. At that point, we discussed the possibility of her producing a new edition, with two important changes.

First, it would be in English. Following the trend in fiction, a text that was already immensely popular in its own language should be given a wider audience. And second, as well as translating it, it would be updated, to reflect new research. And new work is being published all the time, as we discover more and more about the needs, preferences and veterinary care of these fascinating and misunderstood animals.

At this point I became involved as well, to assist with the veterinary aspects of the book, and I am grateful for the opportunity to do so. It is my strongly held belief that, subjected to poor welfare, many other animals run away or turn and attack, but that rabbits, along with other species commonly kept confined, simply stay and suffer in silence. Almost every health problem encountered can be directly traced back to poor husbandry and a lack of preventative health care.

Together with their amazing fecundity, concerns regarding neutering them and misidentification of their correct sex, hundreds of unwanted rabbits are born every day and end up in rescues. Both the degree of suffering endured and the number of rabbits affected are therefore vast.

The solution to this is, I believe, better education for potential owners, breeders, those working in pet shops and rescue establishments, and all those involved in the medical care of the third most popular mammalian pet in the UK. Our aim is that this book will be a useful accompaniment to the increasing range of excellent medical textbooks covering the rabbit, and a resource for educating and inspiring rabbit workers throughout the world.

Whilst I am Veterinary Adviser to The Rabbit Welfare Association & Fund in the UK, I have written here in my personal capacity and, whilst much of this text is in agreement with the policies of the RWAF and other welfare charities, it should not be taken as their official advice.

Richard A. Saunders

Petter. Photo courtesy of Aksel Hunstad, Norway

Introduction

Twelve years ago, I walked around at a shopping centre, peeking into various stores, and by chance ended up in a pet shop. I left with a tiny rabbit.

As many others would have, I felt pity for the small and quite nervous animal. He was left all alone, even though at his young age he should have been together with his mother and siblings. I couldn't leave him, and so Petter came home with me. I had been raised with dogs, cats and rats, but had no experience with rabbits. I also had no idea how hard it would be to obtain correct knowledge on the species, but most of all I was seduced by the little gentleman. His distinct personality and presence gave me an insight into the species' potential, we attained a close relationship, and he is the reason why I ended up educating rabbit people.

Rabbits are the third most common domestic animals after dogs and cats. They are also one of the most misunderstood and underrated animals. Unfortunately, a lot of rabbits suffer due to lack of and incorrect knowledge amongst both pet owners and veterinarians, and rabbits' health and behavioural needs are rarely met.

The rabbit's history as livestock is probably why many still are kept in unsuitable hutches with a diet and lifestyle that are unsuitable for the rabbit's health and welfare. In addition, rabbits are a species that fill numerous different roles worldwide. They are hated as pests and loved as pets. They are bunnies in cartoons and celebrated in children books, they are bred and kept for their flesh and fur, and they are widely used as both laboratory animals and for entertainment. Regardless of how we classify them, all these rabbits have the same needs, and whatever category we put them in, they will suffer from poor husbandry and enjoy well-adapted living conditions.

Most illnesses that affect domesticated rabbits are a direct or indirect result of suboptimal nutrition, and this was sadly also the case with Petter. I followed the advice of various pet stores and listened to marketing from the pet food industry, which resulted in a constipated and acutely ill rabbit. At 2 years of age, his digestion could no longer handle this unsuitable diet, and he was suddenly incapable of moving and close to death. Visits to different veterinary clinics, where they lacked both knowledge and equipment to treat rabbits shocked me, but after steadily feeding Petter with liquids over the next few days, he pulled through. This was the start of the Norwegian edition of *Rabbit Behaviour, Health and Care, Den Store Kaninboka*.

How could it be that it was so hard to get correct information about such a common animal? Why did anyone produce unhealthy food that was both sold and recommended in shops one trusted and would think had better knowledge? Why couldn't the veterinarians treat rabbits, and why did so many people keep their companion rabbit outside in a hutch?

If there was anything Petter had taught me, it was how smart, sensitive, humorous, curious and happy rabbits can be. As a free-range house rabbit he lived like most cats and dogs, and he remained harmonious and affectionate his whole life. However, everyone who came to visit was surprised that a rabbit was roaming free in the house; they had never heard of such a thing before. They were also surprised at how confident, outgoing and fun he was, since most people had the opinion that rabbits were boring animals, just sitting in a cage. They said he was like a dog. I said he was like a rabbit who was allowed to be a rabbit.

I sought to increase knowledge and understanding of the rabbit in all areas that could influence the species' welfare, and so I began to study and provide knowledge to others with an interest in rabbits. I wanted more people to experience a rewarding life with their companion animal as well, and the web resource http://www.kanin.org was established in 2006. It was the first forum for companion rabbits in Norway, and the need for knowledge seemed to be enormous. Through dialogue and guidance of rabbit owners for several years, I obtained a solid foundation for knowing what kind of information people were looking for, what they needed to know and what was important to convey. Through practical experience in the Norwegian animal charity, Dyrebeskyttelsen, I acquired, in addition to the scientific knowledge, valuable experience in terms of dealing with different rabbits and various challenges.

Focusing on what is best for each individual rabbit, I have studied different major approaches and attempted to provide as comprehensive an understanding of the species as possible. With a degree in psychology, I was in particularly interested in behaviour and learning, and concentrated on understanding natural rabbit behaviour in order to offer them suitable living conditions in a domestic setting. Widespread problems could easily be solved when taking their health and prey behaviour into consideration.

In the introduction to *Den Store Kaninboka*, I was hoping to increase knowledge on rabbits and change attitudes towards the misunderstood species in all Scandinavia. I am happy to notice that it has been a popular book, which succeeds as intended; readers are exposed to another view and approach towards rabbits. They are then able to understand their behaviour, health, social needs and welfare, to see them as individuals, which in turn helps them to provide for those needs and to improve welfare of the population, reducing the incidence of neglect and abandonment.

There had never been a welfare- and knowledge-oriented rabbit society in Scandinavia, so in September 2013, I established the Norwegian Rabbit Association, Norges Kaninforening. The Norwegian Rabbit Association is an organization that seeks to increase rabbit knowledge for all involved with the species. We are working to improve rabbit welfare by providing information on rabbit care, in addition to improving the level of knowledge and awareness in veterinarians, pet stores, government and relevant control authority and organizations. I am happy and proud to say that we have already influenced and written the Norwegian Food Safety Authority's brochures on rabbits, an important step in standarizing good rabbit knowledge. We work across the country and are also involved in international issues where rabbits are concerned.

As an author with a relatively small national language, I am excited and thrilled about the opportunity to publish a book in English. People and rabbits face similar challenges worldwide, and I am appreciative of the possibility to reach out to even more readers. With assistance from Richard Saunders, we have made the necessary adaption for the international market, and I am beyond gratitude for his help and support. He has read through all the text, made invaluable comments and additions, and probably struggled with my occasionally creative and bad spelling.

At the time of writing I am living with my four beloved house rabbits: Harald, Even, Melis and Mandel. Very attentive readers might find pictures of them in the book. Please follow me and my rabbits at Instagram@ maritemiliebunny.

Rabbit Behaviour, Health and Care has a Facebook site, which is also the case for *Den Store Kaninboka*. I strive to be available at both, so see you there!

Marit Emilie Buseth
www.rabbit-behaviour-health-and-care.com
www.denstorekaninboka.no
www.norgeskaninforening.no
Oslo, Norway, 10 October 2014

Acknowledgements

I want to thank all those who in various ways have made it possible to create this book.

First of all I will thank Sarah Hulbert for finding me and making it possible to publish this book at CABI. I would also like to thank Alexandra Lainsbury for excellent assistance and feedback, and Lauren Povey for extraordinary help with the design and patience with a picky author.

I will also be eternally grateful to Richard Saunders for being a rabbit guru and helping me out with all kinds of medical information and adjustments for the UK and US markets. I could never have done this without you.

I am grateful to Burgess Excel for sponsoring colour pictures in this book.

Special thanks goes to Robert Søvik for doing the dishes.

I am also grateful to my Scandinavian readers and the rabbit environment we have established online. It has been educational, inspiring and a pleasure to keep such a close contact with all of you.

I am also grateful that Hege Johansen, Nina Lukman Høyrup, Linn Krogstad, Anne Jacobsen, Titti Mjaaland Skår, Gunn Eliassen, Helene Oldeide and Gitte Stormly have been helping me with the forum, especially while working on this book.

A special thank you goes to the rabbits who have shared their stories and pictures.

And last but not least I would like to thank Petter, Harald, Melis, Even, Mandel, Ekorn, Bajas, Furry, Hans, Helene, Håvard, Tøffen, Tom, Tilda, Mari, Melissa, Sigrid Sporveier, Gråtass, M-gjengen, Håkon, Erling, Terje, Brage Bjølsen, Turid, Dixie, Dexter, Lotte, Wall E, Eva, Kalle, Lillemor, Lars, Trulte, Gustav, Pia, Tjorven, Dina, Guro, Emma, Goliat, Murre, Trygve Strømsberg, Humle and all the other rabbits that have inspired me to write this book.

Mandel was abandoned and hiding under a car in Oslo. She was just a couple of months old, skinny and tired. The author heard of this and rescued the frightened rabbit. After neutering, she moved into the author's house as the fourth family rabbit. Photo courtesy of Marit Emilie Buseth, Norway

Pika. Photo courtesy of Ken Kitamura, Toronto, Canada

1

The Origins and Development of Rabbits

The History of Rabbits

Imagine that you are a rabbit. If you happen to end up in an average home, you will probably spend most of your time in a hutch without any company and without the possibility to move about. The inability to hop or burrow will most likely feel frustrating, not to mention the loneliness of being by yourself all day and night with nothing to do but sit, eat and sleep. If, on the other hand, you were a wild rabbit, you would probably be a member of a larger colony living in tunnel systems beneath the ground, and much of your time would be spent grooming, nibbling and playing in the twilight hours.

Sassy and Rufus. Photo courtesy of Hedda Sveum Ødegårdsstuen, Norway

Few people have the opportunity to supply their pet rabbits with such an abundance of land, grass and playmates. There are still, however, a great deal of things that can be done to provide one's pet rabbit with a more satisfying life than those cooped up the traditional way in garages, gardens and basements.

Approximately 3000 years ago the Phoenicians discovered the Iberian Peninsula. In historical antiquity Phoenicia was a merchant maritime culture whose seafarers travelled throughout the Mediterranean area. When, during one of these expeditions, they discovered the area dividing the Mediterranean from the Atlantic Ocean, their attention was also drawn to a tiny, previously unknown animal species. The Phoenicians were overwhelmed by these small lively creatures, who populated most of this new land, digging holes, dancing at dusk and eating grass. They decided to call the country Shepam-Im, meaning the Land of the Rabbits. The Latin translation is Hispania, or Spain as we know it today.

The Phoenicians brought some rabbits home, and the species gradually spread around the Mediterranean area. To have regular access to meat, the Roman army had for several years kept hares in adapted fenced areas, called leporaroa, and to protect the animals from predators, large enclosures were often built with a roof.

Gradually the Romans began to make use of the rabbits as well. The hare-like animals Phoenicians had introduced a few hundred years earlier were thus becoming farmed animals. The Romans understood that the rabbits needed different living conditions, since, unlike hares, they are tunnel dwellers. Around 100 BC they began to expand the enclosures down into the earth, to give the rabbits the opportunity to dig and consequently express their natural behaviour and, in particular, to breed. By the year AD 230 large numbers of rabbits lived in captivity in Italy, and it is believed that the ensuing wild colonies around the Mediterranean Sea were established by rabbits that had escaped such enclosures.

Run rabbit run! Photo courtesy of Sharyn East, USA

It was medieval French monks who laid the foundation for today's domestication of rabbits. Since the flesh of some young rabbits was not considered to be meat, the monks had permission to eat them, even during religious fasts. To ensure a constant supply of this legal meat the farming became more intensive and the rabbits were kept in cages. Somewhere between AD 500 and 1000 the monks started selective breeding, and rabbits with different fur and meat quality were consequently produced. In the 1500s they also began to experiment with different colours and patterns.

Hated as pests, loved as pets

It is considered that the Italians were the first to keep house rabbits. During the Renaissance, it became especially popular with the female gentry to have these new companion animals, and many developed close and affectionate relationships with their rabbits. Grand funerals were held for beloved rabbits. Otherwise, it was most common to keep rabbits exclusively for their meat and fur, and the monks' husbandry spread slowly throughout parts of Europe.

Rabbits both escaped and were released for hunting purposes, and those who managed to survive formed colonies in rural areas in several countries. Rabbits were introduced to Britain as well, and are today the UK's most expensive invasive species. Britain's estimated 40 million rabbits leads to enormous financial costs, including damage to crops, businesses and infrastructure, in addition to damaging the British wildlife. The costs of controlling the invasive non-native species are expected to rise, and rabbits are thus hated as pests or loved as pets.

Rabbits as companion animals long remained a rarity. Until the mid-1800s they were bred mostly as livestock, valued for their meat and their fur. Around 1850, however, an interest in shape and colour began to emerge, and it became more common to breed rabbits in order to compete in constructed breeds and other make-believe criteria. In particular, breeds such as 'Belgian Hares' and 'Dutch' ensured an increased interest in rabbits as companion animals.

During the 1900s a number of different rabbit breeds were designed and established, both as show material and pets. In the 1970s, three different lops were introduced to the USA: the French Lop, Mini Lop and Holland Lop. With their adorable appearance and relatively good temperament, people wanted to keep them as companion rabbits within

Gråtass and Popcorn. Photo courtesy of Tonje Engen, Norway

the house, which brought about an increased interest in rabbits' health and their nutritional, behavioural and social needs. Organizations that wanted to develop and disseminate information about the species were created in the 1980s and 1990s and have been invaluable in terms of raising the standards of this often misunderstood animal. The House Rabbit Society (http://www.rabbit.org) in the USA and The Rabbit Welfare Association & Fund (http://www.rabbitwelfare.co.uk) in the UK are two of the groups that were formed and which have always been a driving force to make professionals, as well as private individuals, become more aware of the needs of rabbits. People became more interested in spending time and money on their beloved animals, and veterinary education and various clinics have had to meet customers' increased expectations and requirements.

Unfortunately, Scandinavia, along with many other countries, lags far behind in this respect. A lack of priorities from the various educational institutions, and little knowledge among the general population, has meant that most of the Nordic rabbits today still are unable to take advantage of the information actually available on the species. Attitudes towards rabbits in these and other countries have mainly been derived and influenced by traditional large rabbit-breeding systems. However, their somewhat old-fashioned and outdated views are not transferable to those who want to take care of the species' physiological, biological, psychological and social needs, and the first forum and group for companion rabbits in Norway was established as late as 2006. The website http://www.kanin.org was set up to be a counterbalance to the existing forum, which was more production oriented, with no knowledge or interest for welfare. In 2013, the author also founded Norges Kaninforening, the Norwegian Rabbit Association, which seeks to improve the welfare of rabbits by providing information about good rabbit husbandry to everyone living with rabbits, as well as helping to increase the level of knowledge to veterinarians, pet stores, governments, educational institutions and various organizations.

The author with two of her rabbits, Even and Harald. Photo courtesy of Siv Dolmen, Norway

Australia, myxomatosis and rabbit (viral) haemorrhagic disease

Moving animals out of their natural environments can lead to serious problems. The devastating consequences that followed the importation of rabbits to Australia are probably the best known example of environmental impacts to which such an entry into a new fauna can lead. The species was introduced to the continent in 1787, when the First Fleet brought several prisoners, officers, wives, children, and apparently some rabbits from Great Britain to Australia. The first British colony in Australia was about to be established.

Rabbits were mainly kept for their flesh, and even though some must have escaped from their cages and warrens, the rabbit population was not considered to be a problem for the first few decades. The original native predators seemed to be more effective as hunters and natural controllers than later imported carnivores, such as foxes and cats, and can probably be one of the explanations why the rabbit population remained low. In addition, most of them were descendants of domesticated rabbits and therefore not sufficiently camouflaged with wild-type colours.

However, the current infestation had its origin with the release of 24 wild rabbits, imported to Thomas Austin for hunting purposes in 1859. At least they were wild looking, but it is believed that some of them actually were grey domesticated rabbits. However, the climatic conditions were perfect for rapid growth and in a short time the population multiplied. Other farmers released their rabbits as well, and colonies spread across Australia at about 130 km/year. In 1926 there were 10 billion rabbits on the continent, and the impact they had on the environment was disastrous. The excessive grazing by the rabbits caused loss of land through soil erosion, destroyed vegetation, and led to loss of pasture and consequently numerous abandoned farms. Rabbits had significantly altered the botanical composition and fauna on the continent, and were also known to be the most essential factor in species loss. Destroyed habitat caused the extinction of both predators and other preys.[1]

Different control measures had been tested and tried since 1901, but nothing seemed to stop the invasive species. Conventional methods such as shooting, destroying warrens, poisoning and hunting did not seem to be very effective, and neither was the rabbit-proof fence that was built around 1900 to protect areas that were still rabbit free. As a result, myxomatosis was deliberately introduced as a biological control agent.

Myxomatosis seemed effective and killed 99% of the population between 1952 and 1954. However, the remaining rabbits showed some resistant to the virus and were able to reproduce and recover. They recolonized old warrens and spread further into new parts of the continent. As with infection in many geographical areas, mortality rates have dropped from approximately 90% down to 25% since the disease originally arrived.

Viral haemorrhagic disease, also known as rabbit calicivirus, or rabbit (viral) haemorrhagic disease (RVHD), is a rapid onset, and frequently fatal, disease of rabbits. RVHD emerged in China in 1984 and killed millions of domesticated rabbits there, and hit mainland Australia in 1995, after previously being confined to a small island off the coast. There is some controversy as to how RVHD has spread throughout Australia, with both natural infection and deliberate dissemination of the infection, as biological control vector, being likely. The virus is very resistant to deterioration in the environment, and has spread throughout the world via inanimate objects.

RVHD was reported in Europe and America in the following years, and arrived in the UK in 1992. The disease became endemic in the wild European species, though it did not seem to affect any North American native rabbits or hares, such as cottontails and jackrabbits. However, periodic outbreaks in domesticated rabbits are a great concern around the world (read more about myxomatosis, RVHD and vaccinations in Chapter 5).

The tremendous cost associated with rabbit infestation amounts to millions of dollars each year, and rabbits obviously have a bad reputation in Australia. Rabbits are known as pests and not pets, and this is probably the reason why people with companion rabbits and veterinarians in Australia request knowledge of the species to such an extent. Rabbits are even illegal to keep in some states, but are known as the most popular illegal pet.

To be considered as a pest has also led to insufficient care being taken of production animals. As producers of fur and flesh, rabbits must pay the price for the lack of knowledge and missing focus on welfare as a despised species.

USA

Europeans who settled in America would also hunt rabbits and introduced the European wild rabbit to South America in the mid-18th century. The few rabbits multiplied at tremendous speed, and in the 1930s they were an established species in the south.

There are several native rabbit species in North America, but all domesticated rabbits are descendants of the European wild rabbit.

Myxomatosis is prevalent amongst domesticated rabbits in the UK and is enzootic in some regions of the USA, in particular California, where the 'California' strain is one of the more virulent strains of this disease. Infection peaks in the hottest/wettest months, when mosquito populations are at their highest, and dawn and dusk when mosquitoes are active are also the most likely periods in which rabbits will be infected.

The presence of the native *Silvilagus* species, who are not affected severely by the virus, acts as a reservoir of infection for domestic rabbits.

RVHD has been noted in the USA in a series of isolated outbreaks in farms and other commercial rabbitries over the past decade or so. The first reported outbreak was March 2000 in Iowa, and a common theme in this and other outbreaks (Indiana, Utah and New York) has been the inability of authorities to trace the source of infection. This is likely to reflect the persistent nature of the virus in the environment and the ease, therefore, of spreading it over long distances on inanimate objects.

Spain and Portugal

Contrary to the devastating consequences that rabbits have caused in Australia, they are not considered as pests in the Iberian Peninsula. One does not see such an enormous increase in population and subsequent extent of damage in the rabbit's country of origin, probably because of natural controls, such as predators, climate and tailored vegetation.

Rabbits are actually considered to be a vulnerable species in Spain, and the population suffered a huge decline of about 71% between 1973 and 1993, with a further decline of 49% in the period 1980–1990, mainly due to the arrival of myxomatosis and RVHD, but other factors such as overhunting and climate variations also seem to have contributed to this dramatic decline across these countries.[2] As a consequence, the viral diseases seem to be a far more effective method to eradicate rabbits in their native landscape rather than in areas and climates to which the species was introduced later.[3]

The rabbit's native countries of origin have thus the opposite problem to Australia, as they are trying to stabilize and contain healthy populations. Rabbits are a keystone species in the Iberian fauna and flora, and are the prey for at least 29 predators, including the endangered carnivore lynx. As is the case in Australia and Spain, we thus see that either importation or reduction of species has consequences for the entire ecosystem.

Big in Japan

In Japan, it is relatively common to have free-range house rabbits, and knowledge, food and equipment are pretty good. If not living with a rabbit, one might stay with the species in special coffee houses in Tokyo, which have emerged for people who either live in too cramped conditions or for other reasons are unable to live with a companion animal.

The European Rabbit

The European rabbit (*Oryctolagus cuniculis*) is known for digging underground networks of burrows, called warrens, where they live together in colonies. The tunnels can be very long and sophisticated, with a number of entrances and separate enclosures for every inhabitant. Rabbits spend most of their time in their burrows, except when they are out to graze or exercise.

With enemies both above and below ground, they must be cautious and alert. Therefore, they

Trulte digging in the garden. Photo courtesy of Marit Emilie Buseth, Norway

often graze together in the safety of a group, so that one can always be on guard.

Whilst one rabbit sits upright, stretches his neck and points his ears, the others can graze and run around freely. Being part of such a group results in a sense of security for the individual rabbit: vigilant behaviour is shared between members of the group, enhancing security whilst giving each individual more time to eat.

Cooperation, combined with intricate tunnel systems, ensures that rabbits also have the greatest possible security underground. The dwelling consists of a network of tunnels bound together by different passageways. The passages are often narrow, measuring only 15 cm in circumference, enough room for only one rabbit at a time. These are joined at intervals by 'meeting places' measuring up to 40 cm. We still do not know how they manage to pass each other, or how it is decided who will back up, yield and other such traffic rules, but a generally peaceful coexistence tells us that this system apparently functions well.

The extensive passageways in the tunnel system are designed to confuse any trespassers who might manage to make their way in, and ensures the quickest possible escape route for all members should they need to flee at a second's notice. Rabbits are well oriented with their own burrows and must have a finely developed mental map of their habitat and its arrangement.

Unlike many of their enemies, rabbits can hop vertically. This enables them to jump through exits in the roof and disappear in a blink of an eye. The confused intruders often find themselves unable to navigate their way within the rabbit's home and, finding the den suddenly abandoned, are forced to slink their way out again, empty handed.

Rabbits live in large colonies that are further divided into groups of two to eight animals. A well-developed social system regulates their living arrangements and leads to a generally peaceful coexistence, even though aggressive competition and fighting occurs. A dominant male reigns, and the other rabbits have different positions in the hierarchy.

Those high on the social ladder enjoy the benefits in terms of access to attractive partners and desirable areas to make a nest; it is therefore of importance to be a rabbit of rank.

Rabbits are gregarious creatures. They sleep together, eat together, play together and groom each other. Who gives and receives the most care is dependent on the individual's social ranking, but whatever position, they enjoy each other's company and form close bonds with selected partners.

Despite their social behaviour and the need to be with fellow members of their species, rabbits are colonial animals that still require some time to themselves. Each rabbit therefore has private rooms at their disposal.

Rabbits are often highly territorial animals, defending the area considered to be theirs. If an uninvited rabbit visits another animal's burrows, or foreign rabbits enter the colony's marked boundaries, then one must act. Such defence rarely takes the form of fights, but intruders must either earn entry into the group or be chased away. (Read more about necessary precautions for social housing in Chapter 4.)

With no real opportunity to protect themselves from predators by making an attack, they must immediately respond to anything that may indicate danger and escape. They alert others by thumping their hind legs on the ground. This alarm will effectively notify the rabbits of danger, giving them a head start.

Harald sleeping in his Cottontail Cottage. Photo courtesy of Marit Emilie Buseth, Norway

Illustration courtesy of Nils Erik Werenskiold, Norway

Lagomorpha – The Rabbit's Relatives

Rabbits are members of the taxonomic order Lagomorpha. Lagomorphs are further divided into two categories, the Ochotonidae and the Leporidae families.

The Ochotonidae family consists only of pikas, whereas the Leporidae family encapsulates both rabbits and hares of different types. All these animals are classified in the same order as a result of a common set of teeth.

All lagomorphs have six incisors. In addition to the four visible front teeth, pikas, hares and rabbits have an extra set of short peg-teeth located behind the upper incisors. Rodents, on the other hand, are characterized by the fact that they only have four incisors. The largest order of mammals is Rodentia, with 2777 species; however, rabbits do not belong to the Rodentia. Rabbits are often mistakenly thought to be rodents, but despite their continuously growing incisors, fur and cute faces, they are not related. Rodents are not even necessarily vegetarians, while the lagomorphs are solely dependent on grazing on fibre-rich food.

Ochotonidae family – pikas

Pikas, or so-called American 'rock rabbits', have small egg-shaped bodies, rounded ears, short legs and no visible tail. Their body length varies from 16 to 21 cm and they weigh between 75 and 290 g, depending on the species. They are not only

Origins and Development of Rabbits

	Order Lagomorpha					
	Family Ochotonidae Pikas	Family Leporidae Rabbits and Hares				
Genus	*Ochotona* pikas	*Lepus* hares	*Oryctolagus* rabbits	*Sylvilagus* cottontails	*Brachylagus* pygmy rabbit	+ 6 other genera
Species	Silver, *Ochotona argentata* Turkestan red pika, *Ochotona rutila* + 27 other species	Arctic hare, *Lepus arcticus* Mountain hare, *Lepus timidus* Black jackrabbit, *Lepus insularis* European hare, *Lepus europaeus* + 28 other species	European rabbit, *Oryctolagus cuniculus*	Eastern cottontail, *Sylvilagus floridanus* Swamp rabbit, *Sylvilagus aquaticus* Marsh rabbit, *Sylvilagus palustris* Brush rabbit, *Sylvilagus bachmani* + 13 other species	Pygmy rabbit, *Brachylagus idahoensis*	

Family tree for the Lagomorpha (from Hoffman, R.S. and Smith, A.T. (2005) Order Lagomorpha. In: Wilson, D.M. and Reeder, D.M. (eds) *Mammal Species of the World*, 3rd edn. Johns Hopkins University Press, pp. 185–193)

smaller but also look different to other lagomorphs. The American species resemble rabbits in the sense that they are social and live in groups, but instead of digging underground tunnels they live together in the cavities found in rocky areas, called talus. Like rabbits, they are colonial and have their private territory within the colony.

Most of the tiny species live in the alpine regions of the western USA and south-western Canada. Like rabbits, they feed on grasses and herbs, but since food is difficult to come by in the alpine environment during winter, they have developed a system where they cut, sun-dry and store vegetation for later use in so called 'hay piles'. If the food ration is threatened by bad weather, they will move these 'hay piles' to a safer place for storage.

Unlike the relatively silent rabbits, pikas have a significant vocal repertoire and communicate with the help of various peculiar short squeaks. They call and whistle to each other, and the vocal abilities are especially useful when the colony is grazing. Like rabbits, they often graze together in the safety of a herd so that one always can be on guard. An alert pika alarms the others of sudden danger, and the shy creatures can disappear in an instant.

Pika. Photo courtesy of Phil Smith, Canada

Leporidae family – rabbits and hares

There are over 80 different species of hares and rabbits. The species in the Leporidae family consist of several hares and a number of rabbits, which all share common features. Their furry tails, elongated ears and hind legs make them different from the previously mentioned pikas, and their family name Leporidae simply means animals resembling 'lepus', the Latin name for hare.

One of the species in the Leporidae family is *Oryctolagus cuniculus*. This scientific name for

All our domesticated rabbits are descendants of *Oryctolagus cuniculis*. Jessie, Bambino and HairyYet. Photo courtesy of Helene Hauglien, Wien

the European rabbit was adopted as late as 1874, and is thus a relatively new term in biological classification. *Oryctolagus* is Greek, meaning hare-like diggers, whilst *cuniculus* is the Latin word for underground passages. Our domesticated rabbit's scientific name is therefore 'hare-like animals that dig underground passages'.

All our domesticated rabbits are descendants of *Oryctolagus cuniculis*. Various breeds are therefore the same animal, just with a different exterior. Regardless of their coat, whether they are large or small, they still have the same nutritional and social needs, the same behaviour and a critical need to move about.

Species in the Leporidae family are adapted to variable environments, temperatures and conditions. Some hares live in polar regions, while others live in the African savannahs. Different types of rabbits live in specific locations throughout the world: in the mountains, in the desert, on the plains or in swamps.

Members of the Leporidae also vary in size and appearance. The smallest pygmy rabbit, *Brachylagus idahoensis*, is found in North America. It is only 25–29 cm from snout to tail and weighs about 300 g. On the other hand, the biggest European hare, *Lepus europaeus*, is about 50–76 cm long and weighs 2.5–5 kg.

Common to all the different species of Leporidae is the fact that they are herbivores, or plant-eaters. Their entire feeding strategy and behaviour is adapted towards a sustenance based on grass, and their specialized digestion makes it possible to utilize such a nutrient-poor diet. Like our domesticated rabbits, they are dependent on nourishment derived from various types of grass, herbs and leaves.

Leporids are also dependent on their ability to escape predators, usually by running in a zigzag pattern to confuse their enemies. They are designed for rapid movements with their light bodyweight and powerful hind legs, have large eyes and an almost 360° field of view, movable ears and an excellent sense of smell.

Sylvilagus – cottontail rabbits

Sylvilagus is one of the genera in the Leporidae family. It consists of 17 species and is found only in America. Different cottontails constitute the group of animals, and they vary in size from the smallest leporid in the world, the pygmy rabbit (*Brachylagus idahoensis*), to the largest member of the genus, the swamp rabbit (*Sylvilagus aquaticus*).[4]

Cottontail rabbit. Photo courtesy of Andy Purivance, USA

The Eastern cottontail, *Sylvilagus floridanus*, is confusingly similar to the European wild rabbit. They look the same, with their grey-brown fur, large hind legs, long ears and a fluffy tail, but there are some peculiarities that distinguish the two species.

The species does not live in social communities in the same way as the European wild rabbit. As with most cottontails, they live in nests called forms and have no need to collaborate in the painstaking task of digging underground burrows as did our domesticated rabbit's predecessors. However, they are still highly territorial and guard their nest above ground. As Eastern cottontails do not give birth to their young in the protection of underground tunnels, the babies are born with a very fine coat. However, they are blind for 4–7 days and do not begin to move out of the nest until they are 12–16 days. The mothers nurse their babies twice a day but are otherwise absent, as with other lagomorphs (see Chapter 12).

The marsh rabbit, *Sylvilagus palustris*, is another tiny cottontail that lives by the coast in the south-eastern USA. Unlike most other rabbits, they are excellent swimmers and always live in regions with access to water. Their feet have less fur than other rabbits, probably an adaptation to allow them to swim more effectively, and they thus seem to take to the water when surprised by a predator.

They weigh about 1–1.2 kg and have small ears and short legs. Crawling on all fours, they have a rather strange gait compared to other rabbits. Placing each foot down alternately, step by step, their walk is almost cat like, probably because it is difficult to jump in the marshy areas in which they live. They are capable of jumping, just like any other rabbit, but seem to prefer walking in the dense thickets.

There are several subspecies of the aquatic marsh rabbit. One of them is the endangered Low Keys marsh rabbit, *Sylvilagus palustris hefneri*, named after Playboy founder Hugh Hefner.

Lepus – hares

For centuries it was thought that hares were solitary animals with limited or no interaction whatsoever. They do sleep alone in their private forms during the day; however, at dusk they often travel to a feeding place, where they seem to prefer the company of others. Even though they do not sit as close together as European rabbits, they do benefit from grazing in the safety of others. Like other lagomorphs, hares feed in a constant state of alertness, and the more hares looking for danger, the more time each individual can spend on grazing.

Hares do not have an equally structured priority of order as in European rabbits, although they still have social rules, and some kind of hierarchy exists amongst males. However, they appear to be far less territorial than the aforementioned species; they do not scent-mark objects or individuals and do not seem to have a well-defined area to protect. Since they normally have enough food, aggressive and territorial behaviour is only necessary during mating in the breeding season.[5]

Mountain hare. Photo courtesy of Steve Gardner, UK

Mountain hare, *Lepus timidus*. Photo courtesy of Steve Gardner, UK

Mountain hare, *Lepus timidus*. Photo courtesy of Steve Gardner, UK

One of the reasons why European rabbits live in larger and more structured groups than hares may be due to their need for underground living. An adequate tunnel system is much easier to dig when you have access to a large number of construction workers. Difficulties in obtaining sufficient land may also force the rabbits to live closer together and in larger colonies.

To protect themselves, rabbits are dependent on a network of corridors beneath the ground. Hares, on the other hand, do not have the same need to hide for their security. Their physical build makes it possible for them to outrun the problem, and they can flee very quickly over large distances.

They are also well camouflaged due to their customized colours, which can be very helpful whilst

Mountain hare. Photo courtesy of Steve Gardner, UK

sleeping in temporary nests above ground. Hares have adapted their camouflage according to their environment. Some species in northernmost America, Canada and Europe can vary from brown in summer to white in winter. The mountain hare, *Lepus timidus*, European hare, *Lepus europaeus*, arctic hare, *Lepus arcticus* and snowshoe hare, *Lepus americanus*, all adjust the colour of their coat according to the seasons' requirements.

Apart from the fact that rabbits wear the same colour coat all year round, hares and rabbits share a similar exterior. Hares generally have longer ears and hind legs, and are on the whole larger than all the different species of rabbits. At the same time, both the rabbit's and the hare's speed is a result of the enormous muscles of the hind legs. These muscles are dependent on an adequate supply of oxygen to perform.

The heart is responsible for the blood circulation, and the more blood being pumped around, the more oxygen and energy there is available to the various muscles in the body. Rabbits have relatively smaller hearts than hares, which affects the two species' endurance. While a rabbit's heart weighs only 0.3% of the total body weight, the heart of a hare contributes 1–1.8% of the body weight. The consequentially more powerful blood flow produces a greater supply of oxygen and energy to the muscles, and explains why the hare is known for sprinting great distances, while rabbits stay closer to their burrows and seek shelter in a nearby tunnel, following short sprints to safety.

Hares have no physical protection in the form of underground burrows, and have therefore adapted to give birth to relatively mature babies. Young hares, known as leverets, are born with long fur and open eyes. They are quite mobile

Jackrabbit. Photo courtesy of Jack Wolf, USA

Origins and Development of Rabbits

Spring and Moses. Photo courtesy of Fat Fluffs Rabbit Rescue, UK

from the moment of birth and have functioning ears and a sense of smell. A mother will wash and groom her young and then leave them on the ground. However, the newborn are so well developed that they can walk away from the birthplace within a few hours. They do not lie down with any siblings, but curl up and stay as quiet as a mouse, several metres apart. The litter of 1–4 leverets is not very well hidden and the young are naturally exposed and vulnerable to predators, but when they are dispersed over a moderately large area like this, it will lessen the chance of the whole litter being eaten.

A corresponding risk analysis is presumably the reason why both hares and rabbits practise absent parenting. If a predator should find a litter, the mother will not be able to defend them. She will therefore avoid undue attention to her offspring and consequently leave them alone as much as possible. She visits the young once or twice a day, nurses them for a couple of minutes, before once again leaving them to themselves.

The mother of the young hares will also attempt to prevent anyone getting wind of her offspring by washing the leveret's genitals after feeding. It is thought that by doing this the mother cleans away the infant's urine to minimize scent clues and help remove all traces. Hares become independent at around 1 month old.

There are a number of different species of hare, and at the time of writing the genera *Lepus* consists of 32 different species. All possess the hare's characteristics, even though they have adapted to different environments. The arctic hare, *Lepus arcticus*, however, is the one who digs the most, usually digging holes under the ground or in the snow to keep warm. It has adapted to the Polar regions with a thick and dense fur, and can consequently live in very cold environments, unlike other members of the Lagomorph order.

The jackrabbit is also a hare, despite the misleading name. There are different varieties, such as the antelope jackrabbit, *Lepus alleni*, which has a very long and slender body, in addition to enormous ears. The large ears act as thermostats so that they can adjust and cope with the heat in the warm regions of America where they live. Other jackrabbits, such as the black-tailed American desert hare, *Lepus californicus*, are also known for their relatively large ears, long legs and bodies, which are quick as lightning. The European hare, *Lepus europaeus*, lives in the northern, central and western parts of Europe. It is this brown hare, as it is also called, that lives in Norway and rest of Scandinavia. They are very timid and can run at 70 km/h.

The mountain hare, *Lepus timidus*, is a smaller version of hare than the European, but is similar in that it changes colour after the seasons, from brown to white. There are also less well-known species such as the desert hare, *Lepus tibetanus*, that lives in north-western China, the Ethiopian hare, *Lepus fagani*, that lives in Ethiopia and Kenya, and the Indian hare, *Lepus nigricollis*, that lives in South Asia.

While hares and cottontails are animals that can both settle and manage on their own, the European rabbit is dependent on large groups for its survival and general wellbeing. This can be transferred to our companion rabbits, which have the same instincts and social needs as their wild predecessors.

Social relations between European rabbits are the most highly developed amongst all species in the order Lagomorpha. Knowing that our domesticated rabbits are their direct descendants makes it sad to think of all the rabbits that live their lives in forced isolation.

Notes

[1] State Government of Victoria (2010) Rabbits and their impact. Available at: http://www.dpi.vic.gov.au/agriculture/pests-diseases-and-weeds/pest-animals/lc0298-rabbits-and-their-impact (accessed 20 January 2012).

[2] Virgós, E., Cabezas-Díaz, S. and Lozano, J. (2006) Is the wild rabbit (*Oryctolagus cuniculus*) a threatened species in Spain? Sociological constraints in the conservation of species. Available at: http://www.escet.urjc.es/biodiversos/publica/Virgos_et_al_2007_Biodivers_Conserv.pdf (accessed 8 April 2013).

[3] Delibes-Mateos, M., Ferreras, P. and Villafuerte, R. (2009) European rabbit population trends and associated factors: a review of the situation in the Iberian Peninsula. *Mammal Review* 39(2), 124–140.

[4] Chapman, J.A. and Flux, J.E.C. (eds) (1991) *Rabbits, Hares and Pikas: Status Survey and Conservation Action Plan*. IUCN, Gland, Switzerland.

[5] McBride, A. (1988) *Rabbits and Hares*. Whittet Books, Suffolk, UK.

Magne and Lill-babs. Photo courtesy of Malin Nilsson Schau, Norway

2

The Rabbit as a Companion Animal

The author with her three rabbits Petter, Harald and Melis. Photo courtesy of Siv Johanne Seglem, Norway

so much more to these subtle beings. The seemingly shy and nervous rabbits are therefore often considered boring and uninteresting, while they are actually active and inquisitive when given the opportunity to be so.

Still, rabbits are not for everyone. To ensure that cohabitation is in the best interest for both humans and animals, it is crucial to have considered all concerns around keeping a companion rabbit before getting one.

> **What to think about before getting a rabbit**
>
> Time perspective
> Allergies
> The rabbit's character (behaviour)
> Costs
> Nutrition
> Living conditions
> The social rabbit
> 'Babysitting' – care for your rabbit when you are going away
> Movement opportunities
> Children and rabbits
> Necessary care
> Choosing a rabbit
> Acquiring a rabbit
> Bringing the rabbit home

Rabbit enthusiasts all over the world are often asked why they choose rabbits for companion animals. People wonder why dedicated animal lovers would not rather live with, for example, a dog or a cat. Most people think of the rabbits they have seen in tiny cages, and do not understand that there is

Around the world, rabbits have come to be popular pets. In Britain, for example, the rabbit is one of the most common companion mammals, only beaten by dogs and cats. At the same time they seem to be one of the most misunderstood and cruelly neglected species we claim to love. In Britain alone, there are up to 2 million pet rabbits, but unfortunately most of them are kept in woefully cramped conditions. An RSPCA survey found that 75% of companion rabbits were being badly treated.

This alarming discovery is also supported by further studies and research. PDSA's PAW Report[1] reveals, as we will see throughout this chapter, that many companion animals in the UK suffer due to lack of knowledge and understanding.

With knowledge of similar conditions and challenges in the USA, Australia and Europe, one can say that these results unfortunately are similar in other countries as well. *Rabbit Behaviour, Health and Care* will seek to increase knowledge and understanding of the rabbit in all areas that will influence the species' welfare.

> 'Stressed. Lonely. Overweight. Bored. Aggressive. Misunderstood… but loved.'
> Quote from Richard Hooker, BVMS (Hons) MRCVS PDSA Director of Veterinary Service, featured in a PDSA PAW Report.[1]

> Both the RSPCA and PDSA's research and advice are based on the five basic welfare needs described in the Animal Welfare Act 2006 (UK).[2] Pet owners are legally obliged to provide the following needs to ensure a happy and healthy companion animal:
>
> 1. **Environment**: the need for a suitable place to live.
> 2. **Diet**: the need for a suitable diet, including fresh clean water.
> 3. **Behaviour**: the need to be able to express normal behaviour.
> 4. **Companionship**: the need to live with, or apart from, other animals.
> 5. **Health**: the need to be protected from pain, suffering, injury and disease.

Jokk and Amos. Photo courtesy of Katarina Valibo, Sweden

> **RSPCA**: The Royal Society for the Prevention of Cruelty to Animals – the UK's largest animal welfare charity. Equivalent to ASPCA, the American Society for the Prevention of Cruelty to Animals.
> **PDSA**: People's Dispensary for Sick Animals. Leading veterinary charity PDSA provides free veterinary care to the sick and injured pets of people in need and promotes responsible pet ownership. The charity operates through a UK-wide network of 50 PetAid hospitals and nearly 380 PetAid practices.
> **PAW Report**: PDSA Animal Wellbeing Report.

Melis enjoys her Maze Haven. Photo courtesy of Marit Emilie Buseth, Norway

Tornerose and Espiranza. Photo courtesy of Lena Mohaugen, Norway

Time Perspective

Buying a rabbit is a long-term commitment. With good care, they can live as long as a cat or a dog, and one should incorporate the animal into any future plans.

Rabbits can live for 8–13 years although the average life expectancy is lower due to incorrect diet and poor husbandry. A thorough study in the Netherlands evaluated the welfare of pet rabbits in Dutch households. The survey revealed that the average lifespan of pet rabbits is 4.2 years, while the potential lifespan is around 13 years. This tremendous distinction can be seen as a result of poor husbandry.[3]

They are totally dependent on daily care, regular checks, vaccinations and a proper environment for all of their life. If you are not prepared to take on this responsibility for such a long time you should consider taking on a rabbit that is already a few years old. Because of the abundance of rabbits, shelters will always have a number of companion animals. Taking care of an adult rabbit is also recommended for children and adolescents who often move away from home during the rabbits' life expectancy anyway. With an even shorter time perspective, one can have a rabbit in foster care for a local shelter.

The Rabbit Welfare Association & Fund, the biggest UK charity dedicated to improving the lives of pet rabbits, found that the number of unwanted rabbits has nearly doubled in recent years. The last meaningful survey conducted by the charity estimated the number of rabbits given up to rescue shelters at around 67,000 per year.

Allergies

Is anyone in your household allergic to rabbits or grass? If someone in the family develops allergic symptoms, the first step is to determine whether the rabbit is to blame. It might be the rabbit's hay, dust in the living room or simply the seasonal trees and plants outside causing these reactions.

The Rabbit as a Companion Animal

If a sensitivity test shows that you are actually allergic to your rabbit, there is no need to rehome it right away. The majority of allergy sufferers can actually live happily with their animal, merely by putting in some extra effort.

First of all, it is important to minimize the trigger to the allergic reactions. Frequent cleaning of the home is necessary if living with a house rabbit. HEPA filters on vacuum cleaners may also be helpful for reducing allergens in the air. Restrict your rabbit's living area, e.g. by forbidding access to the couch or having rabbit-free rooms. It is also crucial to wash your hands after touching and grooming your rabbit.

Second, there are several allergy neutralizers on the market, depending on in which country you live. Petal Cleanse from Bio-Life is a product that might make life easier for those who are struggling with allergies. It removes allergens and other annoyances that cause allergic symptoms, and is astonishingly effective. It helps to rub Petal Cleanse on the rabbit's fur once a week. The brand also carries similar products such as textile spray, air spray and detergent, to provide further help against allergic reactions. AllerPet and AllerPet/C are also similar liquids that can be regularly applied to the fur and skin. One must also be responsible for regular grooming of the rabbit's fur.

If you are allergic to the hay, it might be worth trying another brand. Some people may react to timothy, while they tolerate oat or other grasses. Try different sorts of hay, minimize the contact and always wash your hands after handling it.

Someone who is allergic can also be medicated with antihistamines, and allergy sufferers should talk with their physician.

The Rabbit's Character

It is important to be aware of the species before choosing a rabbit. Many are surprised and disappointed that the little bunny they bought turns out not to be the cuddly animal they expected. Most rabbits do not like to be carried around or to sit on your lap. They are happiest when allowed to sit on the floor in control of their body. However, it is important to obtain trust from the small animal living in your house, and when you have gained the rabbit's confidence it will charm you by dashing through the apartment, throwing itself around in joy and seeking you out to demand treats. One must also be aware of the fact that there actually is a rabbit in the house, and rabbits tend to dig and gnaw. This should not be seen as problem behaviour, and one should instead be prepared to arrange suitable conditions so that the rabbits can unleash their instincts without sacrificing your quality of life.

Many seem to wonder if rabbits are like cats or dogs. I would say they have traits from both species, while remaining quite distinctive. They can be as social and devoted as a dog, while at the same time be as wilful and stubborn as a cat. The main difference is that rabbits are prey, while dogs and cats are carnivores. Managing and socializing with rabbits is therefore different to training a wolf or rough play with a tiger.

An owners' survey, commissioned by a group of animal welfare organizations including the RSPCA, found that 60% of owners did not know that their pets were intelligent, social creatures that need mental stimulation. However, the inquisitive, dynamic and humorous animal will amaze most people living with a free-range house rabbit.

Molly. Photo courtesy of Emma Almquist, Sweden

Costs

Having rabbits will prove to be far more expensive than you thought. The PAW Report revealed that owners dramatically underestimated the lifetime cost of their rabbits. As many as 99% guessed incorrectly, which means that only 1% had a real understanding of the cost of pet ownership. The lifetime cost figures are based on both initial and ongoing costs and owners should be aware of the need for such investments as neutering, microchipping, vaccinations and necessary accessories.

Consumables like hay, pellets and litter must also be accounted for, and in addition there will be a variety of salads and herbs. Rabbit owners must also have the liquidity to pay for veterinary treatment when necessary, which can be expensive. The report shows that veterinary fees are the most underestimated cost by pet owners in general and we strongly recommend insuring all rabbits. Approximately 94% of rabbits are not insured, which means that only 6% can obtain expensive treatment without giving the owner financial worries.

At the time of writing, estimated overall costs are about £1000 per year for up to 12–13 years.

Nutrition

Make sure you can offer the rabbit hay, grass and water as soon as it arrives in your home. You should also give a recommended amount of supplementary food, such as high-fibre pellets or nuggets. However, be aware that a rapid change in diet might upset their digestive system. Therefore, if possible, offer the same pellets as they are used to and change to a new brand gradually.

Hay and grass are essential for ensuring the health and wellbeing of rabbits. It is therefore of great concern that the majority of rabbit owners asked did not know what the correct diet for their companion animal was. The PAW Report found that 42% of rabbits do not eat sufficient amounts of hay with a further 3% not eating any hay at all. The report revealed a diet disaster for UK rabbits and that many use 'common sense' when deciding what to feed their rabbits. However, this is not sensible at all, as 49% reported that their rabbit diet consisted of rabbit muesli, 88% gave carrots and as many as 10% gave leftovers like cake, toast, cheese, chocolate and biscuits. Compared to other species surveyed, rabbits are offered by far the worst diet and suffer as a result of this.

Read more about nutrition and the importance of a fibrous diet in Chapter 6.

Living Conditions

A rabbit's housing should be planned and arranged before the animal enters its new home. If the rabbit is going to live inside the house, rabbit-proofing of one's home is essential for protecting both rabbits and wires. Offer the rabbit non-slip carpets and enough space to run, a litter tray and places to act as caves or hiding places. A rabbit is not invisible, so be aware that you will see traces of the animal in your house. If you have allergic reactions to straw or hay, or become overwrought at the sight of it all over your floor, you may not be suited to life with a grazer.

If the rabbit is going to live outside, make sure to provide both a large escape-proof run and a dry and safe shelter. Make the living areas big enough for housing at least two rabbits, as one must always offer rabbits appropriate company. Whether the rabbit lives inside or outside, a hutch is never enough!

Rabbits that are used to living indoors must not be moved out during winter or when it is cold outside. If you buy a rabbit from a pet store, it cannot be put directly into a hutch and run in the garden when the temperature is low. It must be several degrees above zero, even at night, so that the temperature change is not too great. A rabbit who is used to pleasant and mild temperatures inside the house will not have developed the required protective winter coat, and the rabbit will freeze and could contract pneumonia. Rapid changes in temperature may also lead to other respiratory diseases.

Research reveals that almost half of the rabbit owners asked did not know that rabbits needed space to exercise. The PAW Report also found that 10% of rabbits, around 150,000 at the time of survey, live in hutches that are so small that they can barely jump two hops; 6% of owners did not think the rabbit needed to go outside its cage; and 16% of rabbits only had access to a run no bigger than the hutch. Based on this, it seems crucial to inform owners about a proper environment.

Theo is fit. Photo courtesy of Hedda Aurbakken, Norway

The Social Rabbit

Many people are not aware that rabbits are incredibly social animals and are happiest in friendly pairs or groups. They suffer from loneliness when caged up alone and prevented from behaving naturally, something which in turn leads to abnormal and stereotyped behaviour. Ensure your rabbit has suitable company and make sure they are both neutered, introduced properly and have a sufficiently large living area.

The PAW Report has revealed that a worrying 67% of owners report that their rabbit lives alone, making this a key area for improvement.

Read more about the social rabbit and necessary precautions for a happy cohabitation in Chapter 9.

Social rabbits – Petter, Melis and Harald. Photo courtesy of Marit Emilie Buseth, Norway

'Babysitting' – Care for Your Rabbit When You Are Going Away

You will probably go away for a weekend now and then, or take an extended vacation from time to time, and you must then organize suitable care for your rabbits. A pair of rabbits may be looked after within their familiar home, while rabbits living alone will obviously need human interaction during the owner's holiday.

Movement Opportunities

Rabbits need plenty of exercise. Those being free range or living in large enclosures regulate this themselves, while those confined to a hutch must be offered at least 4 h outside the cage every day, preferably at times when the rabbit is naturally most awake and active.

Children and Rabbits

Our culture is filled with images of children and rabbits together, but contrary to what many seem to believe, rabbits are not suitable pets for children. Baby rabbits are cute, but they grow up fast. After just a few months they are adults with a strong will and destructive tendencies. They require stimulation through company, exercise and learning; they need to be looked after by someone old enough to understand their needs, and who can also detect subtle symptoms of discomfort.

Ragnhild has learned how to interact with rabbits. They are moving about as they please, and they all enjoy each other's company. Photo courtesy of Aunt Marit Emilie Buseth, Norway

Children may also be adorable, but for a rabbit they can be a source of terror. Typical characteristics of young children, e.g. excited shouting, sudden movement and eager hands, may scare the prey animal and either make it run away or defend itself. Children are often fickle and impatient, while rabbits need stability and peace. Because of this, young children will find it difficult to interact with a rabbit and soon lose interest.

Supervision of children who look after rabbits is crucial. A child will often expect the rabbit to sit happily on their lap to cuddle, whilst this is something very few rabbits do voluntarily. This again leads to the child pressuring the rabbit, nagging and following it around, picking it up and getting frustrated that it is not behaving like a soft toy. The animal is also frustrated, and it is therefore important that a rabbit-experienced adult oversees the child's interaction with the companion animal, teaching the child to handle the rabbit properly and consequently lay the foundation for a healthy and happy life together.

The physically fragile rabbit can easily be injured due to poor handling. Basically, rabbits do not like to be held, so children should not carry them around. Being a prey animal, the rabbit will often resist, kick or fight back with a bite or scratch because they are frightened or in pain, something that results in the child dropping the rabbit on the floor. It isn't uncommon for rabbits to have to deal with pain and injury after such episodes. Typical fall injuries include fractures of the ribs or tooth damage, which may lead to further physical and behavioural issues. These rabbits will often suffer with digestive problems and other diseases as well, since rabbits that are stressed are more likely to become ill.

Children often live in the moment. It is easy to get bored with routines, and many will suddenly 'forget' to give their rabbit necessary care when they would rather play with friends and be out in the evenings, and it is therefore essential that the adults in the family are aware of their responsibilities.

Let the child help with daily care, but overall responsibility lies with the adult. You will never teach your child responsibility by just providing it with an animal. The child will learn by observing its parents, learning how they take care of the companion animal and ensure it has good welfare. Children might have limited liability for the rabbits, but during periods when they get bored of the routines, the adults have to either do the work themselves or ensure that the youngster actually fulfils their obligations. Given that neutering and veterinary costs are rarely covered by the pocket money of a 12-year old, the grown-up will always carry the financial responsibility for the family rabbit.

The Animal Welfare Act 2006[4] states that people need to be 16 years old or older to buy a pet. A parent or guardian of a child is therefore responsible for any animal of which a child is in charge. Anyone who is responsible for an animal has a legal duty to ensure its welfare needs are met.

> Some years ago I visited a family where the man gave me a guided tour of their gorgeous garden. I saw an old hutch in the backyard, something that I of course commented on. He told me, completely unaffected, that his two children, at the age of 8 and 10, used to have a couple of rabbits living there. One day, however, he suddenly found them dead. It appeared that the children had stopped feeding them. He then pointed at some furniture and elaborated about some plants he was going to buy. Acquiring a rabbit must therefore be something the whole family has an interest in.

Necessary Care

Having rabbits is more demanding on resources than is generally perceived. Regular work has to be carried out, and the tradition of having rabbits in hutches that are cleaned once a week is not sufficient if we are to take the rabbits' welfare into consideration. These old and outdated ways of keeping rabbits are used by large livestock breeders and in farming where the focus is to simplify having many animals, and are not a part of modern pet animal ownership. Knowing what it takes to improve the welfare of your rabbits, your companion animals should not be made to live under such conditions.

Daily care

Rabbits are very clean by nature and resent having to stay in dirty environments. Both free-range rabbits and those living in cages or enclosures must have a litter tray available. Rabbits like to keep their living quarters clean and so the litter tray should be changed daily or as often as necessary, dependent on how many boxes the animal has access to, and of course on how many rabbits are using it. Keep it clean and your rabbits' home will stay odour-free. Fresh surroundings will also benefit the animal's health.

One must ensure that good quality hay and fresh water is always available. The water should be replaced twice daily. In addition one must offer the recommended amount of supplementary food.

Rabbits are active animals so they need several hours of exercise each day. If they are free range they will regulate this themselves, while those living in cages must be given sufficient hours outside the hutch, particularly at times when they are naturally most active. Make sure your rabbits have appropriate company as well. If living together in a pair they will not need as much attention from you.

Check your rabbit for signs of injury or illness every day. A daily check-up like this should involve observation of the animal's behaviour. Does the rabbit behave normally? Is it running around as usual or is it sitting hunched up in a corner? Has it eaten the same amount of food as it normally does? Signs of change must be taken seriously and one should always consult a veterinarian if the rabbit is ill.

One should gently go over the rabbit's body, from nose to tail, and look for wounds that might become infected, sudden abscesses, running eyes, drooling, wetness around the nose or on the inside of the front paws, and look for urine stains and smelly droppings that are stuck to the fur around the tail. Unpleasant odour and dirty and damp fur must be taken care of immediately as this will attract flies and might cause flystrike (read more about flystrike on pp. 101–102). During summer and in warm weather it is particularly important to check the rabbit for signs of fly larvae attack.

During the moulting season rabbits need daily brushing.

> **How to hold a rabbit**
>
> Rabbits should never be held by their ears.
> Rabbits should never be held in the scruff of their neck.
> Rabbits should always be held in a firm and secure grip. Use both hands. Hold a hand under the hindquarters while supporting the body with the other. Hold the rabbit snug against your body so it does not feel like it will lose its footing.

Weekly care

A weekly examination is a more thorough variant of the daily check-up. Investigate the rabbit's hind feet. The fur under the feet should be thick, like a carpet, and areas with no hair are a sign of sore hocks, which must be treated properly, although the very 'heel' of the foot has no fur directly growing from it and depends on the fur growing over to cover it from the side for protection.

Front teeth should also be inspected once a week. Healthy rabbits have a tiny overbite, and the upper incisors will barely overlap the lower. A veterinarian must adjust overgrown or damaged teeth immediately.

If the rabbit is living in a cage this must be kept clean. The cage itself can be cleaned thoroughly once a week while the litter tray must be changed more frequently. If not providing a litter tray one must ensure that the hutch is kept clean and dry.

Free-range rabbits only require that you clean the litter tray when necessary, apart from the normal tidying and cleaning of your home.

A rabbit on the stairs. Photo courtesy of Ken Kitamura, Toronto, Canada

Monthly care

Your rabbit's nails will require occasional trimming. How quickly they grow and how often you need to shorten them is dependent on bedding and substrate as well as how much time the rabbit has access to a run or play area and how active they are. Rabbits confined to a cage without the opportunity to run free will usually need to have their nails cut more often than free-range rabbits, but whether they live freely or in a hutch, a frequent check-up is important. A clipped nail should be about 4 mm longer than where the nerve stops. It is easy to see the blood and nerve inside light-coloured nails, and with good lightning it can be visible in dark-coloured nails. A good rule of thumb is to clip the nail level with the fur of the foot.

Many find it difficult to cut rabbits' nails. However, this may be easier with some assistance. One can hold the rabbit gently but firmly, while the other carefully cuts the tip of the nail. It is important not to miss the small inner claw on the front paws.

Claw clippers can be bought at any pet shop or veterinary practice; precise, small and sharp clippers are preferred.

Annual care

Regular checks at the clinic are important to keep your rabbit healthy and happy. All rabbits should be registered with a veterinarian and have their necessary vaccinations. At the time of writing, vaccination is neither necessary nor legal in some countries, i.e. Norway, due to the lack of particular diseases. In other countries it is strongly advisable but not legally required (e.g. UK and USA). Contact your vet for advice if you are unsure what applies in your country.

Preventive health care is neglected in rabbits, as revealed in the PWA Report. As many as 54% of rabbits are not vaccinated and consequently have no protection against potentially fatal diseases, and 44% are not registered with a vet. We hope this book will help rabbits by having more enlightened and responsible owners.

Choosing a Rabbit

Once you are convinced that a rabbit will be a suitable pet for you, you must find out where to obtain an appropriate animal. You also need to consider whether you want a baby or an adult, male or female, one or two.

Baby or adult

Rabbit kits are friendly, cuddly and easy to handle. However, they quickly become hormonal and destructive, and the innocent and cosy rabbit may turn into a proper troublemaker from one day to the next. If one has not had rabbits before or is not prepared to deal with a small, energy-filled terrorist, I always recommend getting a rabbit that is at least a few months and preferably over 1 year old. Then you will have a rabbit that might have calmed down a bit, that might not have as great a need to chew and explore all your furniture and which might cause less frustration. Already neutered grown-ups are also easier to litter-train. Last but not least, you can see their personality and temperament better than in an 8 week old.

Male or female

Many people wonder whether they should have a male or female. Rabbits are individuals, and gender cannot really be used to predict personality. However, one can see the differences between the sexes when hormones affect the animals. Un-neutered does might suffer from pseudopregnancy while in heat, and consequently become more volatile in mood during such periods. This hormonal-related behaviour will cease after neutering, and aggressive tendencies and the need to assert oneself will be reduced for both genders.

Rabbits waiting for homes at Fat Fluffs Rabbit Rescue, UK

Neutering of both genders is recommended, as females are especially vulnerable to diseases if not fixed. Up to 80% of un-neutered does develop uterine cancer from 3 to 4 years of age, a risk that is removed by neutering.[5] The procedure is more costly for females than for males. Rabbits adopted from an animal shelter or rescue may already be neutered and these costs may be avoided.

If you intend on having only one rabbit, it does not matter whether it is a male or a female. However, if acquiring two babies, one must be absolutely sure of their gender, as they quickly become reproductively capable and mature. Please note that employees in pet shops and breeders are not necessarily able to accurately determine the sex of very young rabbits, even if they claim to be able to do so. It is not uncommon to hear about someone who bought what they thought were two females and ended up with five extra kits some months later. One should therefore arrange a veterinary examination of your rabbits in time to protect against unexpected pregnancy and neuter both as soon as possible.

How many rabbits?

Rabbits are naturally sociable and it is strongly recommended to keep them in pairs. However, if for some reason you have a single one, it should be able to live indoors with the rest of the family. Your rabbit's welfare is dependent on interaction, and they may develop abnormal behaviour and suffer from boredom and loneliness if left without appropriate company. Having rabbits in pairs will relieve the owner of the need to provide company to a certain extent.

The easiest way to do this is to adopt two already neutered and bonded friends. However, if you already have a rabbit in your household and want to obtain a friend, make sure that both are neutered prior to their first meeting and take the other necessary precautions.

Read more about social rabbits in Chapter 4.

Breeds

There are many different rabbit breeds, and they were originally selected and bred either because of their physical attributes, fur or for meat. The American Rabbit Breeders Association currently recognizes 47 unique rabbit breeds, the UK has about 50, whilst Scandinavia has 62 approved standards. There may be several varieties within the breeds, such as coat and eye-colour variations, but any of them can make a good companion rabbit.

It does not matter if a rabbit is purebred or not, as long as one is not looking for a show rabbit. Most rabbits in circulation are not standardized anyway, as they are mixed breed. However, all rabbits are equally beautiful, have similar needs and require the same care.

Broadly speaking, the larger breeds tend to have a shorter lifespan than the smaller. Smaller breeds tend to be more nervous and have a greater tendency to 'jumpiness'. Small breeds, such as Polish and Netherland dwarf, may be unsuitable for a household with small children who may wish to pick them up, as jumping out of a person's arms is a major cause of injury. Conversely, the larger breeds and giants may be too big for a young person to pick up. Therefore, if you have small children in the house, do not choose one of the smallest breeds as your companion rabbit. A child will often be more tempted to grab a little rabbit and consequently make it more nervous. Adopt a larger rabbit instead as they often seem to be calmer and not as tempting to carry around.

Rabbit breeds can be divided into five categories, and the mixtures will naturally have recognizable ancestry and look fairly similar. There are various dwarf breeds, small breeds, medium breeds, large breeds, giant breeds, as well as breeds with deviant/diverging fur texture. Both the rabbit's adult size and future care should be taken into account when choosing a breed.

Larger breeds require more provision of space and may struggle with certain configurations of runs, tunnels and climbing apparatus. It is therefore helpful to know the estimated adult size of the rabbit(s) to plan the housing for them.

It can be more difficult to predict the future adult size of mixed breed rabbits, but they may be less likely to exhibit certain breed-specific health problems. They may benefit from hybrid vigour[6] or at least the dilution of certain extreme features such as mandibular prognathism,[7] meaning serious malocclusion leading to an undershot jaw.[8]

Long-haired rabbits or wool breeds (e.g. cashmere, angora) require considerable grooming to avoid fur tangling and matting. Lack of care causes great suffering, and these rabbits should be avoided

Pika considers his own standard. Photo courtesy of Ken Kitamura, Toronto, Canada

as companion animals unless one is dedicated to providing the necessary daily grooming. Rabbits with unnaturally long fur have no chance of maintaining a clean and tangle-free coat on their own. Rabbits with normal and short fur require less grooming but still need to be brushed, especially during moult.

Acquiring a Rabbit
Rehoming and rescues

There are an alarming number of abandoned and homeless rabbits in need of new homes around the world, and we therefore strongly recommend contacting a local shelter or rescue before acquiring a companion animal. Numerous rabbits in need will also be privately advertised, because of allergy, illness or other unforeseen events. Many of them are also neglected and forgotten after their owners have grown tired of them. Regardless of the underlying cause, many of these rabbits will be advertised online or on noticeboards in your local pet shop. They are all individuals.

For those in Canada or the USA, please look at http://www.petfinder.com, the number one website for adoptable pets.

Breeders

Those interested in a specific breed may consider contacting a reputable breeder. A quality breeder should know what he/she is doing. They should be able to tell you about the rabbit's parents and further relatives to safeguard against possible inherited dental issues or other diseases. Only social and well-balanced individuals should be bred from, and a serious breeder should also be able to tell you about the different personalities and never raise more than they have a ready market for. Unfortunately, anyone can breed rabbits and call themselves a breeder, simply by starting with a male and female. There is no list of approved or 'registered breeders'. Although Breed societies and governing show organizations exist (for example the British Rabbit Council in the UK), they do not inspect or approve breeders, so be aware that their advice or standards are not necessary welfare oriented and based on science. Always make sure to gather information from reliable sources.

Ask to see how the rabbits are living, including the father. Their living conditions should be clean and spacious enough to exercise and be active in every day. It is also important to know whether the rabbits are socialized and handled gently from an

The Rabbit as a Companion Animal

Taro gets a new permanent home. Photo courtesy of Ken Kitamura, Toronto, Canada

early age, otherwise they may find human contact distressing.

The advantage of buying from a breeder as opposed to a pet shop is that one can see where they come from. You can, by watching their parents, see how big the kits will grow as adults. The rabbits will also avoid the stressful intermediary that a pet shop actually is. Unnecessary changes of environment and diet may lead to an upset digestion and other stress-related ailments, especially at a young age.

Pet shops

Many rabbits are purchased on impulse. People see a cute rabbit in the store and bring it home without knowing what is required for giving the little animal a happy and healthy life. The knowledge and standard in many pet shops has unfortunately been poor, and customers have consequently received inadequate or wrong advice about their new companion animal, although with increased education of vets, owners and pet-shop employees in recent years, this situation will improve.

A recent study investigated the knowledge and attitudes of rabbit owners at the time of purchasing the animal. The respondents revealed a lack of knowledge, especially with respect to the rabbit's diet and social needs, and those who bought the companion on impulse were less willing to neuter the rabbit than if they had thought it through in advance.[9]

The sale of animals over the counter can therefore be said to be one of the main challenges to face when working to raise the status and welfare of the species.

Bringing the rabbit home

When the rabbits have moved into their new home, let them become accustomed to their new surroundings, sounds and smells. Remember that they are a prey animal and naturally cautious and alert. Let them sniff things and become familiar

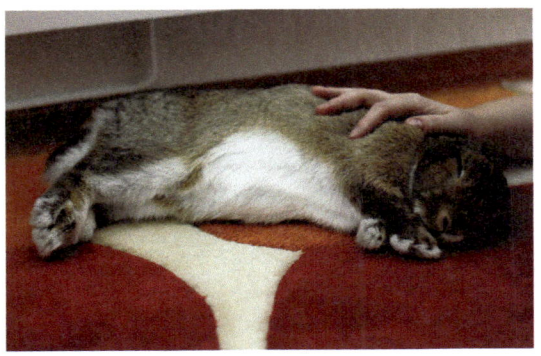

Einar Ensjø was abandoned and observed at a freezing parking lot for several weeks. When Dyrebeskyttelsen, the Norwegian animal charity, finally heard about him, we rescued him at once. He stayed at my office for some days before he was adopted. He is now a content and social, free-range house rabbit. Photo courtesy of Marit Emilie Buseth, Norway

with their new living areas. Do not force yourself on them and keep in mind that they need to figure out the new territory, necessary escape routes and hiding places. Give them time.

The rabbits must learn that you do not pose a threat, and you should therefore avoid picking them up and grabbing them when they try to hide. Sit down on the floor and spend time with your new companion animals on their level. The rabbits will probably come up to you and investigate after a while. It is a perfect opportunity to give them some treats whilst at the same time gently stroking their heads.

If the rabbits are to live in a cage, one should leave the door open, so that they can jump out, either on the floor or in the run, and investigate when they feel like it. If there are children in the house, make sure they do not stress the animals and always provide the rabbits places they can hide and feel safe.

If you have a single rabbit, be aware that they need more attention and company from you.

It may be sufficient that you are in the same room, wandering around, doing what you normally do and sitting on the floor with it from time to time. The rabbit will then get accustomed to your smells, movements and sounds, and become more and more comfortable with you and its new surroundings.

And you will also get to know your rabbit's personality, and hopefully it will be the start of a long and happy cohabitation.

Notes

[1] PDSA (2012) PDSA Animal Wellbeing Report. Available at: https://www.pdsa.org.uk/pet-health-advice/pdsa-animal-wellbeing-report (accessed 9 November 2012).

[2] The Animal Welfare Act (2006) Available at: http://archive.defra.gov.uk/foodfarm/farmanimal/welfare/act/documents/aw-act-2006-memo-101220.pdf (accessed 25 February 2013).

[3] Schepers, F., Koene, P. and Beerda, B. (2009) Welfare assessment in pet rabbits. *Animal Welfare* 18, 477–485.

[4] The Animal Welfare Act (2006) Available at: http://archive.defra.gov.uk/foodfarm/farmanimal/welfare/act/documents/aw-act-2006-memo-101220.pdf (accessed 25 February 2013).

[5] Healtley, J.J. and Smith, A.N. (2004) Spontaneous neoplasms of lagomorphs. *Veterinary Clinics of North America: Exotic Animal Practice* 7, 561–577.

[6] Spencer, R. (2011) Guidelines for entry into meat rabbit production. Available at: http://www.aces.edu/pubs/docs/U/UNP-0080/ (accessed 12 January 2013).

[7] Huang, C., Mi, M. and Vogt, D. (1981) Mandibular prognathism in the rabbit: discrimination between single-locus and multifactoral models of inheritance. *Journal of Heredity* 72(4), 296–298.

[8] Lindsey, J.R. and Fox, R.R. (1994) Inherited diseases and variations. In: Manning, P.J., Ringler, D.H. and Newcomer, C.E. (eds) *The Biology of the Laboratory Rabbit*, 2nd edn. Academic Press Limited, London, pp. 293–320.

[9] Edgar, J.L. and Mullan, S.M. (2011) Knowledge and attitudes of 52 UK pet rabbit owners at the point of sale. *Veterinary Record* 168(13), 353.

Even has learned that this is a smart position for getting treats.
Photo courtesy of Marit Emilie Buseth, Norway

3

Behaviour, Learning and Communication

Gråtass. Photo courtesy of Marit Emilie Buseth, Norway

'Gråtass will be put to sleep tomorrow.' A girl, who cared for horses on a visiting farm, was worried that the manager wanted to euthanize the rabbit and called me. Apparently, it was Gråtass' troublesome behaviour that lay behind the verdict; after all, he bit some of the children who wanted to snuggle up with him.

It turned out that Gråtass was contact-seeking and curious when the girl sat in the pen alone with the rabbit, it was only when he was invaded by a number of children that he responded with this inexcusable behaviour. Not surprisingly, Gråtass did not like being chased, held and dropped by clumsy children's hands, grabbed by the ears and disturbed by a horde of noisy and unpredictable tiny people. He defended himself as best he could, which the manager of the popular visiting farm mistakenly interpreted as aggression, and for that the rabbit had to pay with his life.

The way to deal with unwanted behaviour in rabbits has traditionally been to kill the individual or get rid of it. With cats, dogs and horses, problematic tendencies in the animal have, to a greater extent, been attempted to be corrected through training, learning and adaptation, while in rabbits the problematic behaviour is looked upon as either a defect in the animal or is seen as something quite expected. Unfortunately it is considered normal for a rabbit to sit silently and be listless rather than seeing this as a clear sign of depression, disease, pain or lack of ability to do anything else. In the same way we have come to identify gnawing on the bars as normal, as so many rabbits do this, whereas in reality it is an example of stereotypical behaviour, something of which we will learn more later in the chapter. I will also look at how one can facilitate and resolve some of the challenges that appear to be most common. By raising our understanding of rabbits' needs, of their body language, communication and behaviour, hopefully we will be able to meet their true needs in a more appropriate way.

We have traditionally explained animal behaviour through instinct; however, we now know that both genes and environmental factors play a role in the formation of individuals, both where human beings and animals are concerned.

All animal species learn throughout their whole lives, and the rabbit, being a prey animal, depends on being a good student! Potential danger must be avoided, if not to be eaten! Rabbits learn by use of association and will try to escape from any situation that previously evoked fear or distress. If unable to flee, the rabbit will react by pressing itself against the ground, becoming stiff with fear and making itself invisible. As a last opportunity it may attack.

Gråtass learned that attack was the best defence. If he chased away the kids, they went, and he did

Interaction between rabbit and human – Liselotte and Even. Photo courtesy of Marit Emilie Buseth, Norway

Gråtass and his companion Mindy. Photo courtesy of Marit Emilie Buseth, Norway

not have to be carried around, lost, picked at and bothered. Gråtass was never an aggressive rabbit. He was scared – and he defended himself.

It is highly regrettable that petting zoos, which should teach children about animals and how to interact with them, are not aware of their responsibilities and show us how to treat various animals according to their natural needs. Just as with rabbits, children will also learn from their experiences.

And Gråtass? He moved into a collective with three girls in their twenties. He was offered a litter tray and a hay dispenser in one of the girls' rooms, but also had free access in and out and went happily on a visit to the others in the apartment. Sometimes he asked for herbs in the kitchen while at other times wandered around in the hallway. In the evenings he often jumped up on the couch to his foster mother, and during a week in which she was sick, he lay on the bed and looked after her. Gråtass proved to be very fond of cuddles, but he also knew that he could withdraw if he wanted to. He gained control over his own body, which is essential for a rabbit to feel safe.

Gråtass was adopted, and is currently living with a rabbit companion with access to free roam of the house. And he has, since the day I rescued him, never shown any kind of aggression or frustration.

Rabbits are silent animals. They do not bark when excited or scared, they do not meow when feeling hungry or offended, and they do not display obvious signs of discomfort. Their silent behaviour has developed as a result of the rabbit having enemies on all sides. As a sought-after meal they must always be alert, be as quiet as possible and not attract attention. They communicate primarily through smell, but also have a visual body language and subtle expressions we humans can learn to understand or at least pay attention to.

Abandonment Due to Natural Behaviour

Behaviour problems seem to be the most common reason for the abandonment of animals. Behavioural issues, such as aggression, destructiveness and lack of litter training, are the most reported causes for the surrender of animals to shelters. Undesirable manners, most of which represent the animal's natural behaviours, are usually a result of inadequate living conditions and poor provision for the species' needs. It is therefore important to increase the public's knowledge about animal behaviour, and make them aware of the need to provide for the individual rabbit, cat or dog.

Nature and Nurture

As with all species, rabbit behaviour is also influenced by genetics. The parents should be confident and harmonious animals, as such traits appear to have an impact on the offspring. Genetic conditions in combination with favourable, fortunate experiences is thus a good starting point for obtaining well-balanced rabbits.

However, a less than ideal biological origin does not mean that nervous rabbits cannot be confident and happy animals, as good socialization, safe living conditions and positive experiences will shape all rabbits. However, it is important to be aware that not every rabbit will be comfortable to sit on the lap or be carried around, and that this is something we should respect. A rabbit that does not accept being lifted or carried around is after all just behaving like a rabbit.

Early handling seems to reduce fearfulness towards humans later in life,[1] but either trauma or lack of socialization will likewise affect the rabbit's behaviour. Bad experiences will naturally enough make a rabbit frightened or fearful of new things, but it is thankfully never too late to start the learning process.

Aggression is thus a product of various factors, such as situational context, the animal's previous experiences and biological preconditions.

Aggression in Rabbits

It is no coincidence that rabbits that are perceived as aggressive are usually those living in cages. There are often concerns about so-called problem rabbits attacking owner's hands when putting food in the cage or picking up the rabbit. This is not an

---

Defining aggressive behaviour

Aggression is derived from the Latin word *aggredi* meaning to attack.[2] The definition of aggression has been the subject of considerable confusion, as an animal's behaviour is assessed and defined by an outsider. The animal's behaviour is often considered aggressive when the intention is to harm another, and rabbits are often wrongfully condemned and also suffer from misconceptions in this respect.[3]

Aggression in animals is generally applied to behaviour such as threat and attack, but is an inaccurate and problematic term that often causes confusion. The species is for instance often perceived as angry and aggressive when chasing strange rabbits, when it is rather an expression of the need to establish a hierarchy within social groups. Similarly, a rabbit might attack hands suddenly protruding into their cage, and is often put to sleep due to perceived aggression, when the rabbit is instead defending itself against discomfort and the nearby threat. These characteristics clearly demonstrate biological functions and are part of the rabbit's natural behaviour. The animal's natural response is thus seen as something undesirable, and this is a major concern for companion rabbits. Extensive knowledge of the animal's nature is crucial for determining and defining their behaviour.

Jokk, Amos and Pentti behaving like rabbits in their great run. Photo courtesy of Katarina Vallbo, Sweden

Behaviour, Learning and Communication

angry rabbit but yet another sad example of how misunderstood the species is. Fortunately it is easy to both prevent and correct such behaviour.

Rabbits are primarily interested in surviving. Most of what they are doing is thus motivated by the desire to preserve life. Kits that bounce, run in all directions and throw themselves about in sudden changes of direction can be said to play while at the same time they acquire vital body control. Their comical dance can be understood as an exercise in escaping enemies by being able to change direction quickly and confuse a more linear fox. When rabbits pop straight up into the air they are able to escape predators that hunt their underground burrows. The rabbits build secret exits in the roof of tunnels, which allow them to escape by powerful jumps straight up in the air. The pursuers, however, which are not as skilled at such vertical leaps, run straight ahead and thus end up without lunch.

Like the above-mentioned play activity, aggressive behaviour can also be understood in terms of the rabbit's wish to survive. When a rabbit perceives a situation as threatening, it will defend itself in one of the following ways: it may freeze, flee or attack. The species' first response is to remain unnoticed. Rabbits will try to be invisible and remain perfectly still. However, if the perceived danger is too close, they will try to run away and hide. They will only attack when there is no route of escape available.

The rabbit may feel vulnerable if it is experiencing direct assault against the body, but also if someone is depriving it of important resources such as territory, food, partners and hierarchical position.

> **The four most common causes for aggressive behaviour**[4]
>
> Pain from illness or injury.
> Resource-related aggression.
> Fear-related aggression.
> Redirected aggression.

Pain

Changes in behaviour may be motivated by different ailments, and a rabbit that suddenly becomes aggressive should always be offered a complete health check. In order to exclude any physical ailments, X-rays of the skull, abdomen and spine should be taken. Many so-called aggressive rabbits appear to have severe tooth issues or ailments associated with joint pain, fractures, or abdominal pain. As is the case with humans, other animals will also be more irritable during illness, and rabbits in pain will of course defend themselves against being lifted or having further injury inflicted.

Being a prey animal, a rabbit will always do its best to hide the fact that it is in pain or is injured. With enemies in the air, beneath the ground and lurking behind every bush, the rabbit will try to minimize its chances of being seen and eaten. It will not want to attract unnecessary attention by revealing itself as being weak and therefore the easiest target in the group. This is also why your house rabbit will pretend that it is strong and healthy, even if it has been in pain over a period of time. This can result in both nervous and aggressive behaviour.

It is important to know whether the problem behaviour is due to physical conditions or not. If there is a physical problem present, this must be treated before behaviour problems can be addressed.

Resource-related aggression

Rabbits are dependent on both obtaining and keeping important resources. They must defend their territory, their mates and their position within the hierarchy, as well as their access to food. By being aware of this we can more easily understand why aggression during feeding seems to be a widespread problem, and consequently be able to solve this.

In the wild, rabbits must eat the most nutritious food first in case they suddenly need to escape a fox. It is also apparent that our domesticated rabbits are selective about what they eat, where they scramble and look for specific favourable items in the hay dispenser. They may also attack the hand serving food, and it is in these cases important to improve conditions and react immediately so bad habits do not stick.

Not surprisingly, such food aggression is mainly a problem for rabbits kept in cages, and a good piece of advice is simply to provide the food when the rabbits are free range. Rabbits living in cages or small enclosures must be allowed freedom of movement several hours every day, so if they are given their food while they are out on the floor or in the exercise yard, the problem will be solved.

In cases where one needs to insert the food in a cage with an 'aggressive' rabbit present, you may want to put the bowl of pellets in different places from time to time. However, it is not recommended to be in a position where one must withdraw the hands in fear, since the rabbit learns that such an attack works, and consequently will continue with the successful strategy.

In the wild, there are occasions when rabbits must defend scarce food rations against other rabbits, and if they get a treat, they might take it in the mouth and run off. If you have a multiple rabbit household, it is therefore advisable to sprinkle the daily ration of pellets on the floor, instead of providing it in a bowl. This is to prevent any chasing and squabbles at the dinner plate.

One should also avoid cleaning a cage when the rabbit is present. Respect the rabbit's areas and boundaries and remember that it is in the species' nature to guard its territory.

Resource-related aggression is partly hormonally related. Many people find that rabbits are territorial, often because they guard their cage or the area they consider as theirs. They also protect their position or rank in relation to other rabbits, but also against humans and other animals. Neutering will make these rabbits more manageable, as then hormones will not enhance their behaviour.

> Many rabbits that are kept outside in a cage are unfortunately only visited once a day, usually when they are fed. These rabbits will often show a fear-related aggressive behaviour, due to situations that provoke fear, such as scary hands, loud noises or unfamiliar factors, which rabbit owners misinterpret as aggression during feeding.

Fear-related aggression

The rabbits' defensive strategy is mainly to make themselves invisible or run for safety when in perceived danger. If they cannot escape the threat, they have to resort to plan B, which is to attack.

As is the case with resource-related aggression, behaviour associated with fear is also most common in rabbits living in cages. A common problem is rabbits that are afraid of being lifted. If anyone tries to grab them, they will try to escape by running around within the cage, and when they cannot manage to flee they will eventually attack and bite the person trying to lift them.

Rabbits learn, like other animals, through classical conditioning. Classical conditioning is a learning theory that says that animals learn by using associations. The rabbit sees the impact of various events and thus learns by past experience. This means that if it has an unpleasant experience of being lifted and carried, it will try to avoid this in the future.

An old hutch is standing in the big run. It serves as a hut, and Gråtass is a confident and happy rabbit, living together with his companion rabbit. Photo courtesy of Marit Emilie Buseth, Norway

Behaviour, Learning and Communication

> 'My daughter had a rabbit, but it was totally wild. It was malevolent. It was impossible to approach it, it only attacked…'.
>
> When I tell people about writing this book, they gladly share their own experiences in relation to rabbits. Unfortunately, they are mainly sad stories, most about 'wild' and 'angry' rabbits. These so-called intractable animals were deprived of life and had to spend a life in solitude, often without moving outside the cage at all. After the owners provide me with some information on housing conditions and handling, I can establish the causes of the rabbit's unacceptable behaviour, and give advice on what to do to prevent or improve such conditions.
>
> Few people like to hear that they only had a nervous, scared and misunderstood rabbit, which also probably had physical pain as a result of a sedentary life and poor husbandry. Those who tell me about similar conditions for their existing rabbits are, however, happy to receive simple tips that change the unpleasant situation.
>
> I often get messages about impossible and aggressive rabbits who attack. However, by giving these rabbits a sense of control, freedom of movement, as well as the opportunity to choose whether they want to be groomed by you or not, owners are surprised by the positive result. The aggressive rabbit learns that it has nothing to fear, and becomes a harmonious and pleasant companion animal, and many people by following the advice have been helped to get an enjoyable and more content rabbit. People who previously did not dare to open the cage door tell us about rabbits who have changed behaviour patterns after being allowed to wander around the house at will.

The rabbit has learned that running around in the cage does not help, as he will be picked up regardless, since it is impossible to get away. He therefore begins to bite the hands that try to catch him. This seems to work well, for the hands disappear and the rabbit can stay in safety. He learns that biting is effective, and will respond even quicker the next day to avoid the frightening situation. Eventually he will attack as soon as the door to the cage is opened, and the rabbit is thus now perceived as aggressive.

This association learning means that even neutral stimuli may trigger fear and attack. The rabbit needs to respond to anything that may indicate a predator or an unpleasant situation. When they learn the sequence of events, they will gradually realize a chain of causal relationships and know that it is smart to attack as soon as someone approaches the cage, turns on the light or opens a door. These seemingly innocuous stimuli thus provoke fear in the rabbit because it has learned what they can lead to.

Pain in the body or a sense of insecurity as a result of inappropriate handling will naturally enough make the rabbit unwilling to be lifted. Such a defence is unfortunately often misunderstood, and the rabbit is often noted as a 'problem rabbit', a rabbit that may be abandoned, euthanized or simply left alone in a cage for life. Such a waste of a life and the lack of opportunities to experience pleasure is sad, as it is all about a scared and nervous animal.

Redirected aggression

As with humans, frustration in rabbits might reach a point where it needs an outlet. Frustration as a result of being chased, being in pain or being unable to obtain a certain resource, might result in an agitated rabbit that bites an innocent bystander. This displacement of feelings involves taking out the frustrations, feelings and impulses on people, animals or objects that are within range, or are less threatening than that which caused the frustration itself. It is like when someone has had a really hard day at work and the resultant frustration is taken out on innocent store employees or family. Rather than express the anger in ways that could lead to negative consequences, as in a quarrel with a dominant animal, the rabbit instead expresses its anger towards an individual or object that poses no threat, such as a subordinate rabbit or a carpet.

A rabbit's frustration may be directed against a seemingly neutral counterpart, but the rabbit is not aggressive; sometimes it just needs a release.

Learning for Aggressive and Nervous Rabbits

Rabbits can show aggressive tendencies when they feel pain, and one must therefore rule out physical ailments. It is important to know whether

Harald, king of flops. Photo courtesy of Marit Emilie Buseth, Norway

Help! My rabbit is aggressive!

Lotte was apparently an aggressive rabbit and should therefore be euthanized. She was absolutely impossible and the owner asked for help on a rabbit group online. The rabbit was biting her. What should she do? Several members said that euthanasia was the best solution. 'Poor thing, this was not a pleasant rabbit for you to have,' a helpful member wrote. 'Kill the rabbit and get yourself a new one' was also widespread advice.

No one asked a simple question about the rabbit's living arrangements, under what circumstances she attacked and how the caretaker handled the rabbit. None of those who were responsible for the group intervened, which either testifies to a lack of knowledge or little interest in the species' wellbeing. Nobody found out that the rabbit had lived most of her life in a carrier and was being grabbed when least expected.

Poor and misleading rabbit information seems to be in abundance, and it is important to know where to obtain reliable knowledge and welfare-oriented guidance. Behavioural problems seem to be a recurring reason why rabbits are being handed over to rescues or delivered to euthanasia, which makes this a key area for attention. It is neither challenging nor particularly demanding to improve a rabbit's life, and by following the advice given in this book, 'problem rabbits' will be happier and more harmonious, as was the case with Lotte.

We wanted to help the misunderstood rabbit, and brought her home with us. The scared

Continued

Behaviour, Learning and Communication

Continued.

rabbit could finally be in control. She decided for herself when it was time to leave the carrier, and she explored at her own pace. We didn't grab her but sat on the floor and spoke quietly. She of course showed no aggressive tendencies, and the next day she was already curious about her new humans. She was less afraid of our hands, and we were even allowed to pet her gently on her forehead.

As is the case for many neglected rabbits she was also hormonal, but she was neutered within a few days, and her obvious hormonal behaviour ceased. Being a rabbit, she had defended her territory, which in this case had been a tiny carrier, but now she valued her new freedom and large area, and it didn't take long before she performed her first binky (where the rabbit jumps and turns around in mid-air).

She was also longing for company and was delighted when she finally got two rabbit friends in a wonderful home for life.

Lotte is living happily with her two companion rabbits. Photo courtesy of Marit Emilie Buseth, Norway

---

Dixie and Dexter are free-range house rabbits. The door to the office is closed; otherwise they have a whole floor at their disposal. The rabbits have been present during the renovation of the kitchen, living room and hallway, with no problems, but when they lost access to their own room, there was trouble.

Although the litter tray and hay dispenser were placed in the living room, the loss of their room especially affected Dixie. She scratched at the door and clearly expressed her frustration. She became irritable and began to attack the cats who also lived in the house. She chased her former friends downstairs, would no longer cuddle with her human and was generally in a bad mood.

Not only was the door to the room shut, it was also occupied by another rabbit. Little Pea had to stay there in anticipation of another foster home. Due to lack of space, he had to take over the popular room, something for which Dixie had not given permission. She was obviously frustrated and offended.

One sees in these cases examples of displacement of feelings, or redirected aggression toward something other than what actually triggered the annoyance. The cats were punished for the room being closed.

And what happened when the room finally became available? Yes, Dixie became friendlier, and the cats were once again allowed to move about in the house.

Dixie, the queen of the house. Photo courtesy of Marit Emilie Buseth, Norway

problem behaviour is due to physical conditions, which must be treated before learning can take place. Radiographs of the spine, abdomen and jaw should be included in a veterinary check, as many so-called 'angry' rabbits appear to have serious dental problems or injuries associated with fracture or brittle bones. A rabbit with a sedentary history in a cage will often develop a deformed spine, making it painful to be lifted. That these rabbits object to handling is therefore understandable. Rabbits who have had little opportunity to move about should be offered training by gradually extending the range they have at their disposal. Gradually the muscles and bones become stronger, and the rabbit will be in better shape.

One must never forget that rabbits are prey. They should always have the opportunity to flee, as rabbits have a need to feel that they can be in charge over their own bodies. If they do not have this control, they will easily become nervous. Available hiding places, freedom of movement and to feel in control over their own feet and limbs are thus essential for a rabbit to feel safe.

An efficient and nice tip is simply to let rabbits be free range in the house, either in a room or other confined area. If the base is a rabbit cage, they should be allowed to jump in and out whenever they like, with subsequent control of the situation. Make conditions favourable, so that you are not dependent on lifting them in and out of the cage. If the cage is on the floor, this is easy to achieve by leaving the cage door open.

Harald inspects a human while Petter relaxes.
Photo courtesy of Marit Emilie Buseth, Norway

Spend plenty of time on the floor with your rabbits. Avoid lifting or imposing cuddles on them, but let them sniff around, examining the room and get an overview of the situation. The rabbits may hide at first, but will eventually give in to curiosity and examine you. Gently stroke them on the head while providing a treat, but let them run away if they choose to. Be patient, and the rabbits will learn that they can relax and be confident in your presence. Some might never accept being lifted, but by respecting this you may eventually be allowed to pet a formerly frightened prey.

Treats should only be provided in these training sessions, so that the companion animal learns to associate your presence with something positive. All types of punishment or unintended reinforcement of inappropriate behaviour should be avoided. It is the desirable behaviour that should be rewarded.

In plain language this means that rabbits afraid of being lifted should not have to be picked up and held constantly. Let them, as previously shown, be able to jump in and out of the base themselves, so that they do not develop further unfortunate associations toward humans. This will make them more likely to embrace new positive behaviour patterns.

Always keep in mind that our domesticated rabbits' response pattern is something they derive from their wild predecessors' behaviour. They will react with fear, even in situations where they do not actually have anything to be afraid of, and retraining of the rabbits aims to make them feel more secure.

Once they have previously learned that aggressive behaviour is appropriate and functional, they must be given the opportunity to learn that they have nothing to fear and consequently do not need to attack. The method outlined above shows in a simple way how the rabbit can build up a sense of control and thus learn how to be more secure towards their caretaker and general situations.

Learning in a hutch

It is not advisable to have rabbits living in cages without accompanying runs. However, this is unfortunately still a common way to keep rabbits, and it is naturally more of a challenge to teach these rabbits to feel safe and thus receptive of new knowledge, but one can still get these rabbits to respond more appropriately by means of elementary learning theory.

Rabbits are perfectly capable of understanding the outcome and see the connections of various events, and experiences are therefore essential. As a result of a causal association the rabbit might have learned that seemingly neutral stimuli may trigger fear.

If your rabbit becomes scared when you approach the cage, you should therefore only open the door and put some treats into it – preferably with gloves, so you do not need to pull your hand back if you are attacked. Repeat this until the rabbit is comfortable with you approaching and opening the cage door. Gradually try to pet the rabbit on the back, preferably with a soft brush, while providing some treats. This should be repeated in the same manner until the rabbit is confident and calm with the situation.

Next step might be to carefully lift the rabbit above ground. Gently put the animal down again and reward it at once.

It is important to get a gradual process and not proceed to the next step until the rabbit is confident at the level you are at. If you move forward too quickly the rabbit can be more anxious again. When you finally reach the point where the rabbit tolerates being picked up, you should continue with the amplification of goodies for life.

When Bella and Lillos became free-range house rabbits, they became more confident and outreaching.
Photo courtesy of Marit Emilie Buseth, Norway

Animal charities such as the RSPCA generally receive numbers of animals showing signs of poor husbandry. Many of these show anxiety, but by using positive reinforcement, rabbits seem more secure and at ease after a while.

Mindy was pretty nervous and shaky when she was delivered to her foster home. She had been rescued from a farm with an absence of concern for the little rabbit, and had lived her life in a dirty hutch. She had now been through a check-up, been neutered, and given an enclosure she eagerly watched over. The door was always open so that she was able to walk about as she wanted, but the first day she didn't dare to go outside her new house. She hid in the small cave and guarded her new area.

Her foster parents accepted that they had to be patient, and they never tried to lift her or do anything else that could enhance her nervous behaviour. Instead, they were elated the day they saw a rabbit with pointed ears and long body sneaking around the living room exploring. The humans continued to live as normal and Mindy learned that there was no danger associated with these people walking around. As the weeks went by, she was more often out investigating, she sought out the foster parents who sat on the floor, and got some cuddles and treats. After a few months she jumped up on the couch and begged for grooming, which meant they could also more easily investigate the claws and body.

Mindy took a few months to be more outgoing, and it shows that it is important to be patient. Some rabbits can spend years trying to recover, depending on the trauma, while others are relatively keen and secure after a few days.

Lucy was found abandoned outside in a box. She was in poor condition and very nervous of hands. She could lie down next to her foster parents if they were sitting on the floor, but ran away as soon as they approached her with their hands and tried to touch her. It is likely that she would attack them, if she had not had the opportunity to escape, and the frightened rabbit would thus have been perceived as aggressive rather than nervous. Instead she learned that these hands were not so dangerous after all. She could just flee if they became too intimate and scary. Lucy was a bright student, and after a few days she played happily and gave her foster parents permission to pat her on the back. Today Lucy is a harmonious rabbit, living happily with her rabbit friend.

Other Fear-related Behaviour

Aggression is thus mainly a direct or indirect consequence of fear. Other behaviours may also be motivated by fear, although not necessarily perceived as problematic for the rest of the household. For example, a rabbit that never runs out on to the floor might just be scared. A healthy and contented rabbit will be exploratory and curious, since it is part of their nature to explore different areas. Wild rabbits show a high degree of activity, and if your rabbit does not make use of the opportunity to sneak around, one should make the room more accessible to the rabbit. Carpets on the floor and some cardboard boxes or tables to hide under are often all that is needed to make the rabbit feel that the area is safer.

Single rabbit kits may also tend to be clingy and follow their people around. This can of course be perceived as very cosy, but can at the same time be an expression that the rabbit is timid and needs closeness and a sense of security. Often these rabbits are taken from their mother and siblings too early, and it is therefore important to take good care of them, as they might be afraid of being alone. It is obviously best to let them live with a fellow member of the species, but one should nevertheless spend much time with companion rabbits and teach them to be safe and harmonious animals from a young age.

How Housing Affects Behaviour

Rabbits are often confined in unsuitable hutches, something that can deprive them of the opportunity to behave naturally. Physically, a following frustration might lead to digestive ailments and diseases, while mental stress may result in apathy and depression.

Rabbits unable to be active or to experience pleasure will inevitably spend much of their time sitting still. What else is there to do? Such apathy may well be explained by learned helplessness, a term in psychology explaining when animals learn that they cannot influence the situation. Experiencing a lack of control might lead to the apparent failure to respond in future situations. The rabbit knows it can't escape or fight the scary hands picking it up and will just sit quietly. However, a passive and lethargic rabbit is not a cheerful rabbit. They should be investigating, be curious, active and alert, they should dig and chew, jump and run at full speed and throw themselves on their side in total relaxation.

Anyone living with a free-range rabbit is aware that they have a fuller behavioural repertoire and seem happier than those kept in cages. When using activity levels and observation of natural behaviour as a measure of welfare, scientific studies also confirm that freedom to exercise and be with companions increases the species' welfare.[5]

Do rabbits with access to an outdoor run show a wider variety of behaviour and level of activity than those being kept in a conventional hutch? This was

Dinka is a content and happy house rabbit. Photo courtesy of Katarina Vallbo, Sweden

investigated in a recent study,[6] which clearly showed that rabbits need space. The researchers concluded that the rabbits had far better welfare in the pen or run-around system than in the hutch alone, and that the pen system was the best option to exert their natural behaviour.

Rabbits are thus more active and exert a variety of behaviours when given the possibility. A hutch is not enough!

'A rabbit is in the building!' Amos examines the surroundings and finds Pentti on the other side of the window. Photo courtesy of Katarina Vallbo, Sweden

Normal Captive Behaviour

A rabbit is a rabbit. It may sound obvious, but surprisingly many believe there is something wrong with their rabbits when in fact they only behave like rabbits. They will have a need to dig, whether in a corner of the kitchen, in the garden or in a cardboard box. Rabbits also tend to taste wires, and some help themselves to book covers and mouldings in the process. This is something of which one must be aware, but there are also practical solutions to such challenges, so that both rabbits and humans remain satisfied. (Read more about rabbit-proofing of the home in Chapter 10.)

When the living area is rabbit-proofed and organized, you can observe the species' charming way of life. You will see that they have approximately the same behaviour as their wild predecessors, since domestication has not affected the rabbits' natural needs and behaviour to any extent.

Rabbits spend a lot of their time sleeping and resting, and they are naturally crepuscular. Wild rabbits will therefore mainly be active in early mornings and evenings, while the domesticated rabbits seem to be slightly affected by artificial lights, the sounds of their family and feeding times, and are thus more flexible in terms of when they are awake. It is shown that rabbits living with artificial lighting will be influenced and develop slightly deviant circadian rhythms compared with those who follow the sun.[7] However, they will still have their siesta most of the day, something that allows them to relax while the remainder of the family is at school or at work.

A relaxed rabbit will often rest on its side or belly with hind legs well stretched out. They may also sleep with both the front and rear legs well rolled under the body. A rabbit who falls asleep may twitch, have fluttering eyelids and look as if dreaming. Some snore, others are quiet and some look quite dead. However, all sleep lightly, and if they become aware of a hazard, they are gone in a flash.

Torsan digging in the garden. Photo courtesy of Elise Lier, Norway

Rabbits are creatures of habit. They eat and sleep at roughly the same time every day, they will soon learn when the household gets up in the morning, when they tend to come home from work, that they often sit on the sofa watching TV at night and when it is time for pellets. You will also see that rabbits often prefer to be at various places at different times during the day, which makes it important that they have a sufficiently large living area.

Rabbits will graze about half of the time they are awake. This is crucial for the animals to stay healthy, both physically and mentally. The continuous chewing is required for rabbits to maintain a satisfactory and stimulating life, and is their preferred activity, which also prevents boredom.

Rabbits that spend lots of their time eating hay are less frustrated than rabbits that are mainly being fed concentrated and highly nutritious meals.

Wild rabbits behaving like rabbits. Photo courtesy of Mike Connell, USA (www.txshooter.com)

> **Rabbit habit**
>
> My rabbits are offered pellets around ten o'clock most nights. In good time before serving, actually one hour in advance, they eagerly run around, especially two of them who sit and stare into the closet where food is being stored, and they appear pretty steadfast. Similarly, I hear of other rabbits that also are very concerned to maintain routines.

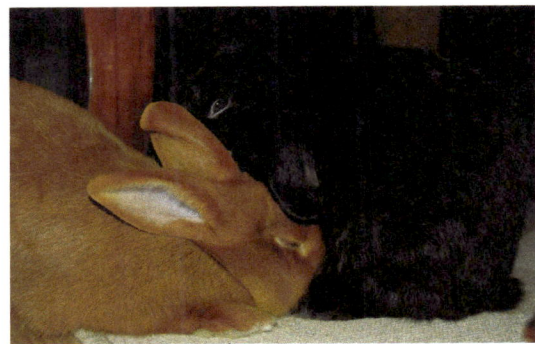

Balder and Frigg are best friends. Photo courtesy of Marit Emilie Buseth, Norway

Rabbits that are offered a diet based on grass and hay exhibit less abnormal and stereotypical behaviour than rabbits who eat large quantities of concentrated supplemental feed.[8]

While hay must be accessible all day, the rabbit's natural behaviour and digestion suggests that the supplementary pellets should be served in the evening.[9] The species will be active at that time, and utilize the supplementary food properly, both with regard to behaviour and physiology. Studies support this contention by showing that evening meals prevent undesirable behaviour, such as chewing on the bars, destroying furniture etc. (Read more about nutrition in Chapter 6.)

Rabbits spend a lot of time keeping their coat in good order. They groom themselves and their partners persistently, and the hygienic animals will not need our assistance to stay clean. Rabbits who live with companions of the same species spend a lot of time washing each other's ears, licking each other on the head or grooming each other's fur. Such mutual care relaxes, and strengthens the bond between, rabbits. They demonstrate affection while grooming each other, and if you receive attention

> **Do not bath your rabbit!**
>
> Rabbits are clean creatures who care for their own and other's fur. One should not bathe or wash rabbits. If they are dirty and suffer from a stinky bottom, one must gently flush and clean the area with lukewarm water, cut off any dirty fur, clean up their litter tray and improve their diet and conditions. A healthy rabbit with a grass- and hay-based diet will have dry stools, which do not cause such problems.

by a rabbit licking your hand, your hair or your forehead, it is a great honour.

Rabbits are naturally active animals. Domesticated rabbits also need to channel this energy, whether through running around, jumping up on a table, digging or chewing. Young rabbits can be especially

Behaviour, Learning and Communication

relentless in this regard. They do not chew on furniture for grinding their teeth, as many seem to believe, but to explore the world through taste and texture, to build up strong muscles of the jaw and because it is fun. The action itself is also carried out with the incisor teeth, not the cheek teeth, and has therefore no effect in keeping tooth length short (read more about grinding movements of teeth in Chapter 5).

Rabbits also like to dig. Instincts help wild rabbits to create underground tunnels and burrows in nature, and this urge to dig, organize and arrange is still held for our companion rabbits. This is evident when your rabbit sits on the parquet floor, digs under the couch and makes holes in the garden or goes mad inside a cardboard box.

A healthy and confident rabbit will often be active. The rabbit's body is designed to keep going, and its curious and playful nature makes them need and enjoy a stimulating environment. Rabbits living in adapted pens along with one or more of their own species have significant welfare benefits compared to those housed alone in cages. They also exhibit less stereotyped and aggressive behaviour when they have better living conditions. By living with other rabbits, they will have the opportunity to satisfy needs and behaviours they cannot possibly find an outlet for in solitude. Nervous rabbits also seem to increase in confidence when supplied with a rabbit partner.

Many observe that the rabbits run around in the morning. Why dash around the apartment or yard? Do they play? Do they exercise in order to stay fit? Do they look forward to having breakfast? It is not easy to say what initiates such behaviour. Probably it is a combination of the above. However, what is certain is that the rabbits seem to be enthusiastic, and socially housed rabbits seem to run in some kind of procession or parade.

At night you will see the corresponding change in activity level. The rabbits will, after the day's rest, need to keep moving. They will run and use their bodies, explore cardboard boxes, jump up on the couch, sneak around the house and occasionally take some well-earned breaks. The rabbit's activities will naturally depend on their preferences and possibilities.

Communication

Rabbits are intelligent animals. They are aware of their surroundings, can distinguish between different

There is a lot going on in here! Photo courtesy of Katarina Vallbo, Sweden

Rabbit Appeasing Pheromone (RAP) may be used to calm down stressed and nervous rabbits. Pheromones are released and seem to affect behaviour in stressed rabbits.[10] At the time of writing, RAP is not available in the UK.

Lars chinning a straw. Photo courtesy of Marit Emilie Buseth, Norway

individuals and will attempt to communicate with the outside world. Communication across species is challenging, and we will never be able to perceive everything an animal expresses. We should, however, be aware of their subtle, visual body language and thus try to understand what they tell us.

Rabbits need rabbits for communicating satisfactorily. They will signal and understand each other primarily by the use of smell. Pheromones, which are chemicals or scent signals secreted by the rabbit, reveal information about an individual's social and reproductive status, and enables them to communicate and influence the behaviour of other rabbits.

Chemo-signalling through the urine and droppings makes it possible to communicate with passers-by of the same species, signalling both gender and ranking within the group. Different

---

Presses itself against the ground: A submissive rabbit will make itself as small as possible so as not to appear threatening. A frightened rabbit may also do the same, but its facial muscles will be tensed and its eyes pushing outwards.

Head thrust forward, ears held backwards and tail raised: The rabbit warns that it will attack. The rabbit will try to sort things out in advance by giving such signals, 'You had best keep away!'

Shakes its head: The rabbit may be irritated; either it senses an unknown odour or has been disturbed, or it might feel it has been groomed by you for long enough.

Pushes with its snout: The rabbit may push on your foot or hand to get your attention and to be petted. In the same manner, when it has had enough, it may push you away. It may also try to tell you that you are in its way.

Nipping or biting: This might mean that it wants you to move. The rabbit may also ask to be left alone, or it may simply be anxious. Baby rabbits will also nip and taste different materials.

Presentation: The rabbit will present itself by sticking its head out at you, laying its chin on the ground, tucking its paws beneath its body and rubbing itself against a person or another rabbit to be petted. The rabbit would like to be stroked on his nose and up to his forehead.

Runs around in circles: One rabbit may run circles around another rabbit, or even your foot, while humming softly. This might be either hormonal or social behaviour, but is most often associated with courting rabbits and disappears therefore after neutering.

Throws itself onto its side: The rabbit is relaxed and quite satisfied.

Exhibits stereotypical behaviour: Sign of stress. Need for improvement in the environment and living arrangement.

Stands on its hind legs: The rabbit is alert, attentive and is getting an overview.

Exposes its back-end: If the rabbit turns his back-end towards you this may mean that you have insulted him.

Stomps on the ground with hind legs: Either the rabbit is scared and is trying to warn the others, or it is displeased and would like to tell you so.

Ears held backwards with opening held into the body: Indicates dissatisfaction. The further back the ears are held, the sadder and more displeased your rabbit is. It should be noted that while sleeping, many rabbits also keep their ears held backwards in the same way.

Ears standing straight up with opening towards you: A satisfied rabbit.

Ears pointing forward, low body and sneaky-like movement: A curious rabbit who is on a discovery mission.

Stands on its hind legs and looks about: An alert rabbit on the lookout.

Licking humans: The rabbit is grooming you – this is a sign of affection.

Moves nose quickly back and forth: Shows degree of interest. Sniffing calmly indicates a relaxed rabbit whereas quicker movements mean that the rabbit is aware.

Sprays its urine: Hormone-related selection/marking of other rabbits as well as territory. This disappears after neutering.

Rubs its chin on objects: The rabbit marks what is considered to be theirs. New areas especially must be explored. Rabbits have scent glands beneath their chin, which enable them to spread information about themselves while 'chinning'.

Jumps and turns around in mid-air: This is called a binky and means the rabbit is happy.

Nips out her own fur and builds a nest: Female rabbits can believe themselves to be pregnant and begin to nest. It is both frustrating and tiresome for the rabbit. Neutering inhibits this behaviour and is recommended.

Tidies up: Many rabbits love to grab and throw away objects that seem to be lying around.

animals have special receptors to note pheromones, and different species will generally not detect or understand each other's signals.

Rabbits will mark their ownership and presence by rubbing against objects with the scent glands under their chins. Rabbits also spread information by leaving business cards in the form of droppings and urine, as well as marking by the use of scent glands under their chin and the anus. Scent is thus used effectively to highlight, warn or provide personal information.

Wild rabbits living together in a colony will make use of a common lavatory. The toilet, which is situated outside their actual burrow, tells other rabbits that the area is occupied and at the same time provides information about its residents.

Males with high position in the hierarchy will make use of several latrines. They will therefore be able to spread their scent over larger areas than the subordinate males. The females seem to spend more time in the toilet, they groom themselves and consequently spread their smell. The males, in particular, will also urinate directly on selected does or partners; an example of hormonal marking that will cease with neutering.

Rabbits also have a visual expression. Different positions on the body and ears will have a specific meaning, and since it can be frustrating for everyone not to be understood, you should try to read the rabbit as well as possible. Rabbits with engineered features, like lop ears, will be less able to communicate by the means of ear position. Rabbits who look more like their wild predecessors, with moving and standing ears, will find it easier to express themselves in this way.

The sound of a rabbit

Rabbits are quiet animals. They have, however, certain sounds they use in their communications. The best known is probably the characteristic thumping. If the animal is in danger or experiencing something scary, they will use the powerful hind legs to stomp the ground and thus warn other rabbits. The others in the colony perceive the warning signal, and can as a consequence run and hide.

Rabbits also express themselves through other sounds:

The rabbit scream: The high and piercing scream signals pain and fear, for example when a rabbit is caught by a predator. The sound is reminiscent of a short infant cry and is surprisingly powerful.

The rabbit grunt/growl: A short grunt, while the rabbit's ears are held back and the animal leaps forward, is a sign of irritation or fear. The tail is usually erect, and it should be regarded as a warning signal. The rabbit may attack.

The rabbit hum: A sound that may resemble an 'oink' or a low, persistent hum, and can be heard when the rabbit is excited. For example, when it circles around your legs as an attempt to courtship.

Grinding teeth: When the rabbit is grinding its teeth, it is a symptom of pain.

Cracking jaws, quick clattering of teeth: The rabbit is enjoying life. Comparable to a cat's purr. Can be heard when grooming the rabbit.

Few people will hear a scream from their rabbits. I have however; when we have rescued rabbits, I have sometimes heard the heart-rending cry. It is traumatic to be caught, and the rabbits are probably afraid of being eaten. Fortunately there is no real danger for these rabbits, and most of them relax after a while.

Behavioural Needs and Welfare

I approach the question of welfare based on whether the animal is able to perform natural behaviours. However, all natural behaviour is neither necessary nor desirable for our domesticated and hopefully safe rabbits. One example is flight reactions due to real or perceived predators, which may cause panic responses if groups are confined together. It is therefore important to distinguish between different types of behaviours and needs.[11]

A need is a requirement for something that is very important. Researchers distinguish between ultimate and behavioural needs. The ultimate needs are those necessary for survival, such as food and water, while behavioural needs are essential for the individual's psychological wellbeing, such as being close to fellow members of the species. Animals have thus both ultimate and behavioural needs that must be met to maintain a satisfying life.

In the wild, ultimate and behavioural needs will generally occur together. A rabbit will both move and interact with others whilst out grazing. However, in captivity we must ensure that both needs are taken care of.

Harald – disapproving rabbit. Photo courtesy of Marit Emilie Buseth, Norway

Hans, Harald's caring father. Photo courtesy of Marit Emilie Buseth, Norway

The animal charity in Norway (Dyrebeskyttelsen) has repeatedly experienced how good the rabbits' warning system works. Rescue operations are far more challenging when there are more rabbits involved, since rabbits constantly make each other aware of what is going on.

This was particularly evident when we had to rescue a rabbit family that had settled in a pile of twigs and stones. Two abandoned rabbits had multiplied, the mother was hit by a car and killed, and we had to save the remainder before the autumn set in.

The kits, who were 3 to 4 weeks old, already seemed to have basic knowledge about rabbit behaviour. They popped into the burrow immediately when a warning was stamped. The father watched over his twig kingdom and three children. Although he was quite confident and contact-seeking, we would not take him before the little ones were safe. Hans, as we called him, often sat 4 or 5 m away from the mound and had an overview of the situation. When we approached one of the kits, he always thumped. Instantly the kits were safely back in the twig pile.

Similarly, one can see that house rabbits will alert each other. Although there is no immediate danger of being attacked by hawks in an apartment, rabbits are careful animals. They will thump following perceived hazards, whether they are alone or in a group, and rabbits living in big farming or breeding systems can similarly trigger fear among their neighbours. If one begins to stomp, this consequently rubs off on others, and finally the whole farm stomps in anxiety.

Behaviour, Learning and Communication

In the case of rabbits, knowledge of the species' social organization and studies on the behaviour of both confined rabbits and their free-ranging compatriots clarifies that companionship and proper social contact is a behavioural need. (Read about social needs of rabbits in Chapter 4.)

The term behavioural need is used to identify behaviour that is essential for the animal for achieving good welfare. If performance of such needs is prevented over an extended period of time it might lead to frustration and ultimately suffering. Experiments have confirmed how important it is to have an outlet for certain behaviour and that this is vital for different species. For example, an experiment examined burrowing behaviour in laboratory mice.[12] It was found that the propensity to dig was equal regardless of whether the mice had access to previously constructed burrows or not. This suggests that the mice need to dig and that the possibility for doing so is important for their welfare.

Sucking behaviour in dairy calves also seems to be internally motivated. A study revealed that calves that were fed from a bucket were still likely to perform their suckling behaviour.[13] Prevented from being with their mothers they were unable to suck milk, but continued to show the behaviour by sucking on other calves or objects. The abnormal sucking behaviour of the calves and the digging mice shows that an animal can be highly motivated to perform behavioural needs. To be constrained from performing these needs could also negatively impact the individual's welfare due to increased stress levels.

Stress

What are the consequences when rabbits have to live in individual small cages that do not provide for their needs? When rabbits lack control and are unable to express their natural behaviour, they are likely to develop abnormal behaviour. A depressed rabbit will seem apathetic, a frustrated rabbit might seem aggressive, and in both cases are likely to use coping methods to manage their lives. Strategies for dealing with such inadequate living conditions may be reduced responsiveness or stereotypical behaviour. Stereotypies are behaviour patterns that are repetitive and have no goal or function, such as bar biting or circling in the cage. Such abnormal stress management indicates reduced welfare, and one has to improve the rabbits' living conditions. There are many people who think it is normal for the rabbit to sit still and look depressed, because most rabbits hutched up look that way. An animal trying to cope with an inappropriate environment is thus the standard for the species' normal behaviour. The abnormal is taken as the normal, and it is important to raise awareness and knowledge about rabbits.

Satisfied rabbits will be peaceful and relaxed. They will jump or sniff around, eat hay, lie stretched out and have harmonious contact with people and other rabbits. However, many are subjected to stress, and it is important to recognize the signs of such behaviour so that one can improve the situation.

There are many who do not know that an animal is also experiencing stress and suffers as a result of it. However, stress in rabbits is often a reaction to the lack of stimuli, poor housing conditions, boredom, travelling, having to attend a show, intimidating surroundings, unfamiliar animals, absence of opportunities to escape, lack of food or water, loneliness or poor ventilation, and other frustrating conditions and poor husbandry in general.

Meaningless chewing of the bars, frantically digging in a corner of a cage, restlessly wandering around and around and other stereotypical behaviour might be the rabbit's way to deal with a difficult situation. When they do not get to live out their natural needs and are prevented from having a normal life, the rabbit tries to cope with a life without stimulation, either by expelling frustration, being aggressive or by being lethargic and depressed.

Personal hygiene and care may cease, or they might pluck out their own fur and groom themselves constantly. These different symptoms of stress-related challenges must be taken seriously. Those responsible for the rabbits must make sure the animals' needs are covered.

Many wonder why rabbits are behaving as they do. They notice rabbits gnawing and tearing the mesh of a cage, and think that this is normal for the species and that there is nothing to worry about. Well, that is not the case. Repetitive and seemingly meaningless behaviour like the above example is considered stereotypical and problematic. Skipping back and forth in a small cage, circling around their own body and plucking fur from themself may be examples of a rabbit who tries to deal with an abnormal situation in captivity. The behaviours can be understood as coping techniques, and are consequently symptoms of a rabbit in an unsatisfactory environment. This both should and can be

Fuzball, a misunderstood rabbit. Photo courtesy of Marit Emilie Buseth, Norway

Fuzball was found in a basket with cushions. Someone had left him in a boutique selling duvets, pillows, bed covers and other equipment, and the terrified rabbit was found by surprised shop assistants.

The Norwegian animal charity (Dyrebeskyttelsen) was contacted and the lost property got a temporary room in a foster care. However, after a while we were told that he was a troublemaker, destroying the house when being home alone. He tore off the wallpaper, made holes in the wall, and was given to someone else who also experienced him as problematic. The rabbit was thus returned to Hege, who led the rescue operation.

Hege's house had for several years served as a more or less voluntary emergency rescue home. Fortunately it was summer, so many of the animals lived outdoors, and with flexibility and commitment, Fuzball also got to stay in her home.

He actually ended up having most of the main floor available, and seemed delighted about his new living situation. Indeed, he had to share the floor with four cats that went in and out as they pleased, but he accepted the cat situation. Fuzball always wanted to be a part of what happened and showed himself as a social and friendly rabbit. He did not eat walls or taste any furniture either, and had obviously benefited from a more stimulating environment.

Fuzball was probably destructive in his former home in virtue of being frustrated. He had, correctly enough, been free-range in a room, but it was quite naked, and without a slip-resistant substrate he had not even had a chance to run and jump. Despite good intentions, the living arrangements had not been good enough. He needed company, the ability to run on anything other than a smooth floor, as well as cardboard boxes and chew toys he could destroy.

Fuzball was considered an impossible rabbit, a destructive rabbit nobody wanted. However, like all rabbits, he just needed a social and stimulating life to be happy.

Fuzball is by the way still a happy boy, living with a rabbit companion.

prevented or solved by providing the animal with better living arrangements.

Rabbits' play

Most species of mammals play.[14] One of the defining features of play is that it is spontaneous, aimless and internally motivated. We recognize the amusing activity when we see a cat who enthusiastically tries to catch a matchbox dangling from a piece of string, or when dogs are play fighting with their caretaker. However, in other animals, it may be more challenging to recognize and determine whether it is play or other behaviour that is observed. Rabbits will not hunt balls or chase rolls of toilet paper, and because of that many people perceive them as boring animals. However, this is a misconception due to lack of knowledge and poor housing, which will cause many rabbits to be secluded and forgotten. One cannot compare prey species with carnivores. Throughout the book, we see that knowledge of the species' behaviour is crucial for understanding the animal, which is also the case for play behaviour.

Rabbit play looks beyond definition. Vigorous body pirouettes and head flicks may look like some kind of convulsion but is actually a rabbit having fun. Running around at high speed or twisting the head and body in opposite directions are examples of rabbits demonstrating enjoyment. Are they playing? I would say so, and so would most people observing such contented individuals. Not all rabbits jump in the air when delighted, some are content with simply an ear flick, but the above-mentioned

Petter, before destruction. Photo courtesy of Jens Petter Salvesen, Norway

Petter and Melis, after destruction. Photo courtesy of Jens Petter Salvesen, Norway

Rabbit play. Photo courtesy of Elise Lier, Norway

behaviour is a prime example of one of the three categories of non-human play:

- *Solitary locomotor play* involves activities such as running, leaping, shaking, twisting and other seemingly erratic movements, and is typical for species subjected to predation. A rabbit having a binky, or simply just a head flick, is thus playing.
- *Objected play* is typical for carnivores, and involves activities that can be related to how the animal acquires food. This means that a predator may use a live animal as a toy, like when a cat chases and tortures a mouse without killing or eating it.
- *Social play* often involves elements from sexual and aggressive behaviour, and the best known example is play fight. Most people are familiar with play-fighting in dogs, both with the caretaker and other members of the species, and because of their outgoing behaviour and close relationship with humans, dogs are known as eager and enthusiastic animals.

We see that rabbits' play is associated with activities such as unpredictable running and torso twists in the air. Because of this, many people find it easier to interact with a cat, dog or even a rat. Their predatory play may include the caretaker to a greater extent than is the case with rabbits. This is likely the reason why a significant number of children become disappointed and bored with their rabbits, and makes this knowledge a key issue to prevent abandonment.

Immature animals seem to engage in play that looks similar to activities they will perform as adults. Based on what animal you are, you will have a certain play behaviour that serves to gain skills that will be needed as a grown-up. Kittens take the hunting licence test when fighting with curtains, dogs are learning how and when to use aggression when play fighting with their caretaker, and rabbits, on the other hand, are practising their abilities to escape predators, when zigzagging or turning around in mid-air. The impressive body control and flexibility rabbits are developing is crucial throughout their entire life. We thus see that play has long-lasting positive effects on the development of animals, but it also provides immediate results, such as confirmation of the social order and physical fitness.

> Some forms of play can simply be understood as evolutionary behaviour, while other play cannot be said to have any purpose beyond having fun,[15] as apparently is the case when one of my rabbits teases me. In the evening, when I have seated myself comfortably in the sofa, Melis sneaks up to a table. She is very well aware that it is forbidden to gnaw on this furniture, and she never tries to otherwise, except when I am going to relax and perhaps watch a movie. When I finally get up to stop her, she runs joyfully and twists in the air. It is as if she is laughing out loud. When I sit back down, she sneaks up once again, and according to Melis, it is just as much fun every time I am forced to get up.

Even goes haywire. Photo courtesy of Marit Emilie Buseth, Norway

Play will only occur in a stress-free environment. A nervous rabbit in pain will rather be on guard than having fun. One should also note that circling in the cage is a stereotypical and abnormal behaviour that should not be confused with carefree play.

Behaviour, Learning and Communication

The presence of play is an indicator of the individual's wellbeing and should therefore be encouraged (see rabbit play at http:www.maritemilie.com).[16]

Absence of play and rebound behaviour

In rescued rabbits I often see individuals who bear the marks of a life with an absence of play and opportunities. Passive postures that characterize depression and apathy are common in these animals, who do not seem to know how to explore, move about, gnaw on materials or use cardboard boxes. For many, it seems to take a long time to learn how to play, while company of a friendly rabbit assistant seems to be inspiring.

Large numbers of rabbits are housed in environments that restrict their ability to perform natural behaviours. Not only will these rabbits show a reduced behavioural diversity, they will experience the extent of boredom and cage frustration as well.[17]

When animals are prevented from practising a natural behaviour pattern for a period of time, they might compensate for the suppressed needs when given the opportunity. Such an increase in the performance of a long awaited activity is called rebound behaviour, and activities such as body pirouettes, leaping and other play behaviour are easily observed in rabbits released from a hutch and offered access to a run on the grass.

A recent study substantiates the theory of rebound behaviour.[18] The effects of spatial restriction were examined, and showed that rabbits had a higher activity level when moved to a larger pen. The rebound effect indicates that when housed in small pens, rabbits are unable to perform normal activities and behaviour, which has a negative effect on their welfare.

Activation and play

Rabbits are grazing animals and will naturally spend much of their awake time searching for something to eat. Making some of the food less available will stimulate this foraging behaviour. Hay must always be readily available, but pellets and treats can advantageously be made more inaccessible, so that the rabbit must strive and spend time to find and ingest nutrition. A grass-based diet is not just important for maintaining good health but is also crucial to prevent boredom. A rabbit that gets a bowl of concentrated and nutritious pellets may only spend half an hour a day eating, while the species in the wild spend most of their time on meals. Encouraging this natural behaviour is therefore important to keep the rabbit entertained.

Rabbits will show interest in toys that can be consumed or contain anything edible. There are a number of different products that are safe and fun to chew on, such as balls, curves and tunnels in unpeeled willow and other suitable material. Cardboard boxes and other toys will also have the charm of novelty, and should be replaced from time to time to ensure variety.

Also remember to have a non-slip surface so that the rabbits can run and twist around.

Activity outside

Rabbits like to be outside in the summer. In countries where the temperature and other conditions make it prudent to have rabbits in the garden, one can create simple exercise areas of compost fences or build more permanent solutions. It is important to note that the enclosures set up of such compost fences are not sufficiently secure and can only be used under supervision. One must also ensure that the rabbits are not placed in direct sunlight, but that they are able to lie in the shade. All enclosures must also contain hideouts. A rabbit that does not have the opportunity to hide will be nervous.

For more information on rabbits outside and necessary preconditions, see Chapter 11.

Rabbit jumping

Some rabbit owners, mostly children and youngsters, enter their animals in agility or rabbit jumping competitions. It is possible to teach a rabbit to run a particular track and force various obstacles along the way, but it is important to note that this is not an unproblematic hobby. A rabbit that lives most of its life in a cage will have a poorly developed bone structure, and thus be prone to bone fracture and other injuries by sudden efforts and the large movements involved in a steeplechase. This applies particularly to rabbits kept in cages outdoors in the winter, when moisture or cold leads to inflammation of joints and additional pain associated with a sedentary life.

Activity is of course great, but it is more important for the rabbit to get regular freedom of movement than attending jumping competitions

Hiding food

A paper bag can keep a rabbit active for a long time. Add some herbs deep into the bag and see if the rabbit will investigate it. Please remember to cut open the carrier-handles, since the rabbit's head can otherwise become stuck.

Pellets can be served in a treat ball. Adjust the opening and let the rabbit work out how to push the ball around so that food drops out.

Add hay and herbs in toilet paper rolls and watch the rabbit throw it around to get hold of the goodies.

Sprinkle the pellets in a box with toilet paper rolls and discover a keen rabbit searching for food.

Hide pellets and herbs around the room or apartment and let the rabbit go hunting. During the night the rabbit will have found all the pieces and be satisfied.

Picnic in a bag. Photo courtesy of Marit Emilie Buseth, Norway

Hunting in toilet rolls. Photo courtesy of Marit Emilie Buseth, Norway

once a week. It is also essential that the rabbit has the physical prerequisites to cope with the challenge. One should also be aware that competitions first and foremost are fun for the owner, and that travelling to and from events with unknown people and animals could lead to stress reactions for a rabbit. Place some obstacles in your garden and stimulate the rabbits by letting them learn how to solve an assignment, but let the rabbit stay in familiar surroundings.

It is also crucial to take the rabbit signals seriously. If a rabbit is not willing to jump, one must respect this. Perhaps there is pain associated with the activity, it may be afraid or simply does not want to be part of the show. Contrary to what many think, jumping over obstacles is not part of the rabbit's natural behaviour. They have powerful legs and are obviously able to jump quite high and long, but they rather get on in life by running under twigs and furniture or digging underground.

On the other hand, there are many rabbits who apparently like to bounce through different paths. These are usually healthy and secure rabbits who have a track out in the garden and have learned that they get a treat after completing the race.

However, rabbits will initiate jumping themselves. They are curious and want to get up in height to keep looking. Many find their favourite place on the couch, a table, a chair, or in Petter's case, on the piano. He jumped up via the piano stool and lay down behind the music stand.

Learning

Rabbits will learn that one event leads to another. The animal sees the impact of various happenings and will thus learn by past experience. This association learning means that a certain stimuli will be followed by a conditioned response; e.g. a

Rabbits need to dig, chew, explore, destroy, clean up and fix. Cardboard boxes are great for such activities and are both an affordable and easy pastime for your rabbit. There are different options, and you can both make and buy exciting castles and houses.

However, regular packaging is still a hit.

Set up a cardboard box, make an opening and let the rabbit devastate it.

Harald inside a box. Photo courtesy of Marit Emilie Buseth, Norway

Benjamin is a healthy and contented rabbit. Photo courtesy of Mia Kristoffersen

rabbit that hides in a corner when someone opens the cage door – the rabbit may have had unpleasant experiences related to people and poor handling, and will accordingly try to avoid being lifted. Seemingly innocuous and neutral stimuli, like approaching the cage, can therefore provoke fear, because the rabbit has learned what it may lead to.

Associating two correlated events is called classical conditioning, which is one of the main theories of how animals learn. Many of our own experiences and reactions are also shaped by the pairing of stimuli. Humans and other animals make these associations all the time, even when we are not aware. Think of situations where a certain smell results in intense emotions. It might remind you of a former boyfriend or a specific episode, and is all because you have made associations, just like the frightened rabbit in the cage.

The other type of learning is called instrumental or operant conditioning, meaning that a rabbit may learn to perform specific responses due to its association with a positive or negative reinforcement. It is similar to the above-mentioned theory, but while classical conditioning is largely passive, operant conditioning is more active in the sense that the animal will influence and affect the situation.

We learn this way every day. When making a mistake you most likely remember it, and consequently choose to do things differently in a future, similar situation. The same holds true when you achieve a positive result. A favourable outcome will make it more likely to perform the same behaviour later, just like rabbits who have learned that they will be rewarded with treats when standing on their hind legs.

We see that rabbits will learn as a result of the consequences of their actions, and this operant conditioning is also the principle behind clicker training.

If you have a rabbit that bites on your legs to get attention, you can simply ignore the rabbit when nagging. If your rabbit does not achieve the desired response, which often will be to cuddle on the nose, the rabbit will learn that it is not appropriate to nibble on toes.

Learning policy

Desired behaviour should be rewarded. Undesirable behaviour should be ignored. Do not punish the rabbit or use other forms of physical correction. One should never slap the rabbit on the nose.

Clicker training

Clicker training derives from the above-mentioned learning theories, and refers to the shaping of behaviour through positive reinforcement. The clicker itself is a small plastic box that makes a sharp, clicking sound when being pushed. The sound is paired with a reward, usually food, and the animal will learn to carry out the desired behaviour when they hear a click.

Eventually, the rabbit will try to get you to click so that it can receive a prize. A rabbit who has learned to jump up on a chair by the use of clicker training might start jumping hopefully up on the furniture to achieve a click.

Clicker training has become extremely popular for all kinds of animals, and is especially helpful when applied to captive wild animals or frightened rabbits. Because there is no need for physical contact, it has been a successful method for teaching terrified animals to become more secure and cooperate with necessary handling.

Behaviour that is rewarded seems to be reinforced, while behaviour that is ignored appears to fade away.

It is quite possible to train the rabbit without using a clicker. These are the same learning techniques that are used when they learn by experience, and it is the aforementioned associations that make my rabbit run towards me when I smack my lips. They associate the sound with the serving of some goodies, since they always hear it before they get pellets or herbs.

Learned helplessness

Many rabbits sit quietly in the cage, clearly depressed and apathetic. Frightened rabbits may respond by biting when someone approaches them, while others rather resign and allow people to pick them up, even though they do not like it. This obvious defensive reaction is a result of learned helplessness, a term in psychology that describes the apparent failure to respond after experiencing a lack of control in previous situations. Learned helplessness occurs when a rabbit is repeatedly subjected to an aversive stimulus that it cannot escape. Eventually, the rabbit will stop resisting, since the animal has learned that there is no possibility of escape from harm or pain. There is no point in trying to improve the situation, even when opportunities to escape are presented. Learned helplessness thus prevents any action and may explain apathy in individuals.

We can see this in rabbits used to being carried around by children, even though they have brittle bones and other painful disorders resulting from rough handling. The rabbits might surrender instead of trying to fight back, due to learned helplessness. We also see this in rabbits that do not know how to walk around when finally given the opportunity. Adapted environment, patience and a rabbit companion will solve the problem.

Tonic immobility

Rabbits can go into a condition where they look almost hypnotized, either when they are laid on the back or handled in some other way. It is a widespread misconception that rabbits are relaxed in such cases when lying immovable on their back in your lap or in the crook of your arm. However, this is not the case. Tonic immobility is a condition certain prey have acquired to escape enemies. It is a defence mechanism to which rabbits habitually revert when they have no other options to escape (read more about tonic immobility on pp. 113–114).

Customs and Practice

Rabbits will demand that you show them suitable respect. Many rabbits are touchy and become offended if you do something unacceptable, as to walk past without grooming their snouts, laugh at them or if you simply have trimmed their claws.

The rabbits can ignore you in several ways, and if a rabbit turns his back-end towards you, while staring into the wall, you might have an insulted rabbit.

Rabbits may forgive you if using bribes in the form of herbs and other goodies, or if you gently stroke them on their head. Rabbits greet and show respect by pushing their noses against each other, and they are therefore very pleased to be patted in the area from snout and up towards the head.

A rabbit who approaches you, lowers the chest and cheek to the ground and stretches his head forward, presents itself and requests to be cuddled. Rabbits can also dig strenuously on your foot to achieve this attention. This may be an attempt to show dominance towards you, as rabbits who are highest on the ladder have an advantage and should be groomed by the submissive members of the group. Some rabbit pairs are not as bound by hierarchy and will give and receive mutual grooming, while in other groups it is more clearly visible who reigns.

Pubertal Rabbits

Pre-teens: Dixie, Dexter and Dolly. Photo courtesy of Hege Johansen, Norway

The entry of hormones can change your innocent and affectionate little rabbit overnight. They are suddenly uncontrollable and mischievous, they are gnawing and digging, and might need to mark themselves with urine and droppings around the house.

Adolescence can occur when the rabbit is about 2 months. When they are 5 to 6 months, all rabbits are considered adults with subsequent hormonal behaviour.

People seem to be surprised when their rabbit no longer behaves like a docile and cuddly little baby. Many rabbits are no longer wanted when this happens, and their owners will either store them in a cage or clear away the 'problem' in other ways. Many of the rabbits that are abandoned and end up in rescues are just hormonal gangsters.

There is plenty of good advice available for coping with life with a teenage rabbit in the house. It is all about attitude. Most people are aware that bringing a puppy into the house involves a lot of work. The little dog will destroy shoes and furniture and relies on training, monitoring and a lot of company. The same applies to a young rabbit.

1. First of all, one should neuter the rabbit as soon as possible. Neutering of both sexes will remove the hormone-related behaviour. However, young rabbits will need to make adjustments and receive good experience to become well-adjusted adults.

2. Let the rabbit dig and chew, but direct their attention to suitable products. If your rabbit digs on your carpet or gnaws at a chair, you should give them a cardboard box they can dig in and offer them some twigs they can taste instead. One cannot remove the rabbit's need to gnaw and dig, but it is easier to prevent frustration and unwanted destruction when showing the rabbit where it is allowed to devastate.

3. Make sure your rabbits can succeed. Let them be in situations where they are unable to do any harm. Punishment works poorly when it comes to raising rabbits. One can provide a limited area in the beginning, so the rabbits might learn to dig in a box instead of a couch. Once they have learned to gnaw on twigs instead of furniture, you can advance and let the rabbits be free-range.

4. Please be patient. The rabbit will calm down when it grows up. The first year, however, is the most challenging, and for that reason I often recommend inexperienced rabbit parents to adopt a grown rabbit.

Many believed that Harald would never adapt and feel secure in captivity. He was the result of two rabbits that were abandoned and multiplied. He and at least his two brothers were born out in a pile of stones on a field, and we rescued him when he was 4 or 5 weeks old. His first meeting with humans was thus when I caught him with a landing net.

In anticipation of capturing the remainder of the kits and parents, I took him home with me. I was not going to keep him, but rather put him together with the rest of the family when we got hold of them. That was not what happened.

He was tiny and clearly the smallest of the kits. The mother had been hit by a car and killed about a week earlier and we were afraid that he might be harmed by poor nourishment. Raised on twigs, nutrition-poor grass, and some rotten fruit thrown right next to their burrow, he was surprisingly fit. He was not skinny, but the coat was stiff, dirty and full of mange.

Continued

Continued.

He seemed, naturally enough, frightened and shocked by the kidnapping, and tried to make himself invisible behind the little house I provided for him. I let him have peace, and was delighted when he ate some grass before sleeping. It was dark and quiet in the bedroom where he lived, while Petter sniffed and patrolled outside the door.

Petter had been an only child for nearly 6 years. Well socialized with people, he went about freely, lay in front of the TV when it was time for the news, sat in the office and worked at night and lay in bed with me at bedtime. I thought it would be difficult for him to accept and understand another rabbit after so many years. It was not.

However, the first days he appeared to be quite ignorant. He noticed that there was another rabbit in his apartment, but did not seem affected in particular. Harald, on the other hand, was thrilled when he became aware of the other rabbit. It was touching to see how excited and happy he was.

The days went past, and I had still not introduced the rabbits to each other, but when Petter slept on the piano during the day I gave Harald free access to the apartment and veranda. He enjoyed himself quite well in the sun and rolled around his own axis. If he was scared or anxious, he ran to me.

After 4 days it was apparently the time to hook up. Petter directed the show, ran over the piano keys, had an intermediate landing at the piano bench and started to jump toward the porch. Since this was a highly unauthorized strategy to introduce rabbits I paid close attention but soon realized that he posed no threat. The rabbits sniffed at each other, Harald showed submissive behaviour, Petter seemed to be pleased, and from that day they stayed together.

Petter was Harald's hero. Wherever he went, Harald accompanied him. Ideally, he would stay close to his best friend all the time, but when Petter thought it was a bit too intimate, it was nice to sleep in a line too. It was obvious that Harald felt more secure because of Petter, who on the other hand, gained a new lease of life. They made each other well.

It has been 5 interesting and instructive years with Harald. Since he was not socialized with humans for the first 5 weeks of life, many rabbit experts thought he would be marked by fear in captivity. However, by providing the most basic criteria, such as neutering, freedom of movement and control over his own body, he has evolved into a well-balanced rabbit. He shows great joy, runs happily around in the house and is a sensitive and kind boy. He is naturally wary of strangers but is easily persuaded with a treat.

After a year Harald also welcomed a newcomer when little Melis was found on a roundabout and eventually moved in with us. I have seen how well the rabbits thrive in groups, and when we lost Petter I was the whole time aware that we had to adopt another rabbit. Fortunately Even showed up, and Harald is now the master of the house.

Notes

[1] Bilkò, A. and Altbâcker, V. (2000) Regular handling early in the nursing period eliminates fear responses toward human beings in wild and domestic rabbits. *Developmental Psychobiology* 36(1), 78–87.

[2] Mills, D.S. (2010) Aggression. In: Mills, D.S. et al. (eds) *The Encyclopedia of Applied Animal Behaviour & Welfare.* CAB International, Wallingford, UK.

[3] Mills, D.S. (2010) Aggressive behaviour. In: Mills, D.S. et al. (eds) *The Encyclopedia of Applied Animal Behaviour & Welfare.* CAB International, Wallingford, UK.

[4] McBride, A. (2009) Help for nervous and aggressive rabbits. *Rabbiting on,* Autumn 2009, s.22–25.

[5] Seaman, S. (2002) Laboratory rabbit housing: an investigation of the social and physical environment. Available at: http://www.ufaw.org.uk/pdf/phhsc-schol1-summary.pdf (accessed 8 April 2013).

[6] Redrobe, S. (2011) Is a hutch enough? A comparision between hutch only, hutch & pen and hutch & runaround systems. The 2011 RWF Conference, Solihull, 29 October 2011. Veterinary Professional Notes.

[7] Nelissen, M. (1975) On the diurnal rhythm of activity of *Oryctolagus cuniculus* (Linne, 1758). *Acta Zool Pathol Antwerp* 61, 3–18.

[8] Berthelsen, H. and Hansen, L.T. (1999) The effect of hay on the behaviour of caged rabbit (*Oryctolagus cuniculus*). *Animal Welfare* 8, 149–157.

[9] Krohn, T.C., Ritskes-Hoitinga, J. and Svendsen, P. (1999) The effects of feeding and housing on the behaviour of the laboratory rabbit. Available at: http://www.la.rsmjournals.com/content/33/2/101.full.pdf (accessed 8 April 2013).

[10] Bouvier, A.C. and Jacquinet, C. (2008) Pheromone in rabbits: preliminary technical results on farm use in

France. Available at: http://world-rabbit-science.com/WRSA-Proceedings/Congress-2008-Verona/Papers/R-Bouvier.pdf (accessed 8 April 2013).

[11] Keeling, L.J., Rushen, J. and Dunchan, J.H. (2011) Understanding animal welfare. In: Appleby, M.C. et al. (eds) *Animal Welfare*, 2nd edn. CAB International, Wallingford, UK.

[12] Sherwin, C.M. et al. (2004) Studies on the motivation for burrowing by laboratory mice. *Applied Animal Behaviour Science* 88, 343–358.

[13] Rushen, J. et al. (2008) *The Welfare of Cattle*. Springer, Dordrecht, the Netherlands, p. 303. Cited in: Keeling, L.J., Rushen, J. and Dunchan, J.H. (2011) Understanding animal welfare. In: Appleby, M.C., Mench, J.A., Olsson, I.A.S. and Hughes, B.O. (eds) *Animal Welfare*, 2nd edn. CAB International, Wallingford, UK, pp. 13–25.

[14] Burghardt, G.M. (2005) *The Genesis of Animal Play*. MIT Press, Cambridge, Massachusetts. Cited in: Pellis, S.M. and Pellis, V.C. Play. In: Mills, D.S. et al. (eds) (2010) *The Encyclopedia of Applied Animal Behaviour & Welfare*. CAB International, Wallingford, UK.

[15] DeMello, M. (2012) Animal behaviour studies and ethology. In: *Animals and Society. An Introduction to Human-Animal Studies*. Colombia University Press, New York, pp. 349–373.

[16] Buseth, M.E. (2013) http://maritemilie.com/2013/07/14/se-friske-og-glade-kaniner-som-leker (accessed 23 July 2013).

[17] Gunn, D. and Morton, D.B. (1995) Inventory of the behavior of New Zealand White rabbits in laboratory cages. *Applied Animal Behaviour Science* 45, 277–292.

[18] Dixon, L.M., Hardiman, J.R. and Cooper, J.J. (2010) The effects of spatial restriction on the behaviour of rabbits (*Oryctolagus cuniculus*). *Journal of Veterinary Behaviour: Clinical Applications and Research* 5(6), 302–308.

Anton and Tina. Photo courtesy of Tina Solicki, Norway

Best friends. Photo courtesy of http://www.eddyrambo.com

4

Social Rabbits

Rabbits are happiest with other rabbits,[1,2] and veterinarians and others involved with the species should be familiar with their social life and encourage owners to keep them in neutered pairs or compatible groups. There are, however, many who believe that the animals will be enemies, as most people have seen or heard of rabbits fighting. However, on the basis of knowledge about rabbits' social system, behaviour and biology, it is possible to provide satisfactory conditions so that they can derive welfare benefits of living with one of their own kind.

> **Necessary conditions for a happy cohabitation**
> - Neutering of both rabbits.
> - Introduction in a neutral place.
> - Sufficiently large and suitably adapted living area.
> - Stability.

Jokk and Amos. Photo courtesy of Katarina Vallbo, Sweden

Neutering

Both rabbits in a pair should be neutered, regardless of their genders. If only one is neutered, the other one will still be affected by hormones, resulting in frustrated behaviour expressed towards the other. This will result in a poor foundation for permanent friendship. Please note that males may be fertile for 3–4 weeks after neutering, and that the hormones may affect behaviour for both sexes some weeks after the procedure. It is therefore recommended to postpone the introduction until this tendency has ceased. (Read more about neutering in Chapter 7.)

Living Conditions

The rabbits have to live in an area that is actually adapted for the species. Cages that are often considered normal for rabbits and the ones that are sold in pet shops are not sufficient. The rabbits must have the ability to choose whether they want to lie together or be apart. They are also dependent on the ability to chase and run from each other to establish and maintain a hierarchy. A hutch is never enough.

Introduction

It is very important that rabbits are brought to and introduced in a neutral area. The species is highly territorial; so they should be introduced in a place neither of them has been in before. The travel both before and after the introduction also appears to be essential.

Bonding

There are different ways to bond rabbits, and consequently different opinions on what is best.

Bella and Lillos. Photo courtesy of Marit Emilie Buseth, Norway

However, many years of practical experience and consultation with hundreds of rabbit owners has led me generally to recommend what I call Method 1 (below), a tried and tested method that is both efficient and often seems to be successful, as long as all the advice is followed.

However, one must be aware that, regardless of method, it is a process that requires patience and supervision. It must also be emphasized that both rabbits should be checked regularly for injuries during the bonding and then daily afterwards.

Method 1

Both rabbits should be transported to the new place in separate or the same travel case and released simultaneously. One must be aware that they will run after each other, hump each other, chase, nip and perhaps tear off some fur. This is normal and an important step in establishing the necessary rank order. Humping on the other's head and back is an attempt to dominate, and this can continue for a few hours or some days. It may look dramatic for the owner, but as long as they do not attack and actually bite each other, they should not be separated in this process.

After a while the rabbits might rest, eat or groom themselves nearby each other. This is a good sign and one may eventually prepare to take them home.

A few rabbits will not chase and ride on each other at all. They might ignore each other instead. Regardless, after the first introduction in a neutral

Stein chased and humped Charlotte for about an hour and a half. This is normal rabbit behaviour when establishing social order, and the rabbits should not be separated. Photo courtesy of Aksel Hunstad, Norway

place the rabbits should be brought home together in the same carrier. The mild stress of the trip will most likely make them seek solace in each other and become closer. Such stressful situations might also be an effective solution if you have rabbit quarrels at home in the future, but moving to neutral ground for a time and then back to old ground is also a good solution for resolving conflicts.

Finally at home the rabbits can be released where they are supposed to live. Make sure they are provided with at least two hiding places, two litter trays, two feeding stations and two water bowls. This is to prevent one of the rabbits being chased away from his water and hay in the bonding process. In addition the rabbits need a sufficiently large living area so that they are able to run and hide.

The rabbits should be supervised for some time, and you might have to sleep on the couch some nights, but when they seem to rest and groom themselves near each other, they normally sort things out.

It is recommended that the process takes place indoors or in an area where you can easily keep track of the animals.

Method 2

Alternatively, some prefer that the rabbits are gradually introduced instead of the quicker method explained above. The idea is that they should be accustomed to each other's smells and become familiar under more careful conditions. The rabbits

Charlotte and Stein have been brought to a neutral place for their first meeting. Photo courtesy of Aksel Hunstad, Norway

are often placed in nearby enclosures, so that they can sniff each other and consequently be more familiar when they first meet. Exchange of litter trays may also help the rabbits to get used to each other's scent. After a while, short and frequent meetings in a neutral territory can be arranged, and the rabbits are separated if there is any sign of tension between them. This should be repeated until the rabbits approve of each other. The rabbits should not be left alone before they are happy to groom each other. The process is estimated to take anything from hours to a couple of months.

It is the author's opinion that this method seems to prolong the process unnecessarily. The rabbits do not get the opportunity to establish an actual hierarchy when separated constantly, and due to the importance of stability this seem crucial.

Combining methods

For the vast majority of rabbits it will be sufficient to follow the first-mentioned method, while for others it might be necessary to try the second or combine the two procedures. If the rabbits have recurring, genuine fights (see box on pp. 63–64) it will be necessary to separate the rabbits, have them in pens next to each other, and let them meet under supervision for a short period.

During the following meetings one should follow the above advice, such as moving to neutral ground for a time and then back to old ground, rearranging the furniture and other steps carried out in the bonding process. Some caretakers become worried and think that they will never succeed; however, by being patient and tolerant, and letting the rabbits chase and hump, most will have the pleasure of observing two rabbit companions after a while.

Alternative methods

If, in spite of the above-mentioned methods, you still have rabbits fighting, there are alternative methods one can try. The use of the territory of other rabbits who have been removed can be a solution for extra-challenging couples. Because the area belongs to another pair or group, the rabbits being bonded hopefully join up to protect each other. Experienced rabbit workers have found this to be helpful for squabbling pairs when all else has failed, but care should be taken to ensure the rabbits will not be exposed to infections and diseases, such as *Encephalitozoon cuniculi* (EC) for example (read about EC on pp. 105–107).

In addition, it is worth mentioning that some choose a process wherein they start with keeping the rabbits in a small area, and then gradually increase the available space. It is the author's opinion that this method prolongs the bonding unnecessarily. Sometimes they squabble as soon as the space is increased, and the space needs to be reduced again. The rabbits' behaviours and needs are restricted during the process, and the method seems to be slow and less stable compared to the others explained above.

I am supervising the bonding of Mille, Mia and Lotte. I am using Method 1, and the rabbits were brought to a new run in the garden. They moved back and forth to the living room inside, where I slept on the couch some nights. Read more about Lotte in the box on pp. 35–36. Photo courtesy of Marit Emilie Buseth, Norway

They never fought, and as long as rabbits have sufficient place to escape, the submissive rabbit will usually lie down like Charlotte or run away like Lotte to prevent confrontation. However, it is crucial to supervise the bonding. Mille, Mia and Lotte became close friends in a couple of days. Photo courtesy of Marit Emilie Buseth, Norway

Is it fighting or not?

The main challenge related to the bonding process seems to be uncertainty as to whether the rabbits are fighting or not. Many owners are also anxious of whether the rabbits are stressed, scared or if they would rather be alone. It is normal to be worried, but it is good to know that rabbits in the majority of cases can handle such a situation, and that they will enjoy the benefits of a future cohabitation.

So, is it fighting or not? In most cases, the activity performed is associated with an establishment of order, such as chasing each other, humping on each other's bodies from all directions and pulling

This is obviously not a scheduled photo, so the image is blurred. I was going to take a picture of Even, when Harald surprisingly attacked. He was ill at the time, and so the sudden aggression was caused by pain. I caught the sudden assault by camera, but intervened and separated the rabbits at once. It did not take long before they were great friends again, and they have been ever since. Read more about this episode in the box on p. 69, second column. Photo courtesy of Marit Emilie Buseth, Norway

Continued

Social Rabbits

Continued.

out bits of fur. It is necessary that the rabbits are able to test their relative strength, including the attempt to dominate by humping and chasing the other. If other precautions, such as neutering, a proper introduction at a neutral place and sufficient living conditions are taken care of, there is no need to separate the rabbits in such a situation.

However, if rabbits truly fight there will be no doubt, and one should intervene immediately, as they can hurt each other badly. In a fight they attack each other's head or genital area, they might hang together and roll across the floor and consequently end up hurting each other if not removed. Always check for wounds and provide treatment if necessary. If the rabbits are biting and attached to each other one must take care when separating them, as one could accidentally tear the rabbits and cause further injury to them. If one is attentive and prepared and if the aforementioned precautions have been followed, a separation is rarely necessary.

It has been a common belief that fighting prevents future bonding, but if you follow the steps carried out in the bonding process above, this is hardly likely. Be patient, adjust the conditions and allow any hormones to calm down before trying again.

Combination of Rabbits

Many people wonder what pairs of rabbits work best. Rabbits are individuals, and some get along better than others, but most rabbits will tolerate each other if the aforementioned recommendations are followed.

Unfamiliar rabbits

Female and male

If one already has a rabbit and wishes to acquire a companion, it is most common to select one of the opposite sex. They seem to be most tolerant of each other, but must be neutered for obvious reasons. This is generally held to be the most compatible and stable pairing and is therefore recommended.

Two females

Many does live together, and they are generally tolerant of each other. However, hormones will affect un-neutered females, and since they must defend imaginary nest sites during heat, a less stable hierarchy may lead to frictions in the group and consequently quarrels. Neutering will reduce this hormone-controlled behaviour.

Two males

Previously, this has been an unthinkable combination, as many seem to have thought that two males would fight each other no matter what. However, after they are neutered and other precautions are taken care of, two males can be as good friends as other mixed groups. Although, it seems to be easier to achieve a long term relationship if they have grown up together or are neutered as soon as their testicles descend.

Probably due to fighting behaviour between un-neutered or only recently neutered males, many still avoid this combination.

Siblings

Many people buy two rabbits from the same litter in the belief that they are of the same sex. One should always get a veterinarian to double-check their gender, as employees in the pet shop or breeders may have difficulties determining their sex at the age of only 5 or 6 weeks, which unfortunately is a common age at which to sell rabbits. However, the kits should be with their mother and siblings until they are at least 8 weeks old.

Because of the early sexual maturation of the males, the mother should be neutered when her kits are 11 weeks of age, or the males should be neutered or separated from their mother until this is done.

Females mature later, but when they reach 15–16 weeks of age one must ensure that they are not kept with un-neutered males.

It is the authors' opinion that males should be neutered at 11–12 weeks of age and females at 16 weeks of age, this to prevent hormonally related squabbling amongst males in addition to making sure no one is impregnating another rabbit.

Due to the rabbits' need for stability, both rabbits in a bonded pair should be brought to the clinic when one or both is having an appointment with the vet. The rabbits should be kept in the same kennel whilst there, so that they can seek comfort and support in each other and maintain their bond.

Regardless of family relationships, make sure that all the advice in this chapter is followed to ensure a well-functioning cohabitation.

Why bonding is necessary. E6, an un-neutered male, was put directly into another rabbit's hutch. It ended up like this. Please follow the advice in the text to avoid this. Photo courtesy of Marit Emilie Buseth, Norway

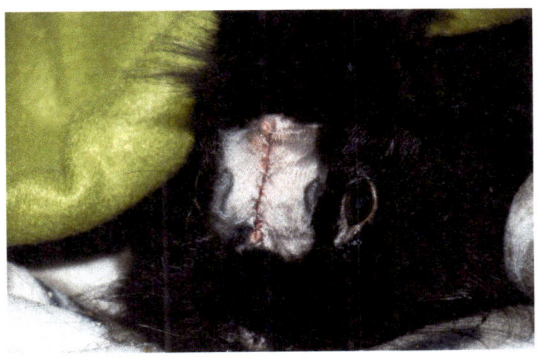

Why bonding is necessary. Guro sneaked into another pair's enclosure and was attacked by the other female. She was immediately taken to the veterinarian, the fencing was rabbit-proofed and Guro is now comfortable with her partner again. Photo courtesy of Aksel Hunstad, Norway

Håkon lived in a hutch for many years. He was sitting alone for six cold winters and six hot summers, with nothing to do but sleep and eat. Regardless of whether he was freezing or caught in the heat, he couldn't do anything but turn around or look out through the mesh.

His owners complained about the boring and impossible rabbit, and never let him out of his hutch. After 6 years they were tired of him and wanted to give him away.

I found someone willing to give him a second chance, and Håkon was picked up by his rescuers and brought to his new home.

He was not used to moving about and went cautiously around in his new room. He was clumsy and seemed amazed that he could jump. It was also an adventure to stretch properly. All the years in a hutch that was too small for the rabbit to lie down at full length had caused him to develop a distinctive technique for stretching. First he stretched his front paws while yawning, while the rest of his body was crunched up. Then it was his hind legs that could be stretched, while the front legs remained still. Fortunately, after a couple of weeks he became comfortable lying stretched out on the floor. He became more exploratory every day, and his health seemed to improve. However, another curious rabbit was lurking around in the living room. Sam was about to get a friend.

Both the males were neutered, and their caretakers waited a few weeks before the bonding was going to take place. Meanwhile, Håkon stayed in the bedroom, while Sam had the rest of the apartment at his disposal.

After the hormones had calmed down, both rabbits were transported to a neutral place. During the car trip they sat in separate carriers, but were released simultaneously at the chosen location.

Both sneaked out of their carriers and carefully inspected the area, well aware that there was another rabbit in the room. After a curious nose-to-nose confrontation, Håkon ran off, Sam reached him and started humping him on his head. Håkon lay submissively down and allowed Sam to climb on to him on both his front and rear-end, before they proceeded to chase each other.

When they decided to have a break, they found a private hiding spot, groomed their own fur and relaxed for a while, before they had to continue the quarrel. By the next confrontation it was Håkon's turn to hump on Sam, but when they were finally tired of chasing they sat down close to each other eating pellets. The caretakers were relieved, as this is a fortunate situation that may take hours or even weeks to achieve.

During the ensuing journey home, they were placed in the same carrier.

Continued

Continued.

Affected by both the intense situation and car trip, they sat nicely together before they again were released at home. The living room had previously been Sam's territory, and he chased Håkon around to tell him just that.

However, it was not long before they rested and groomed themselves near each other, and both humans and rabbits finally seemed somewhat relaxed. To ensure further progression and to supervise the situation, one of the caretakers slept in the living room for the following couple of nights. However, the rabbits quickly agreed and came to rest.

After 6 years in total solitude Håkon had finally gained a friend. His level of activity increased day by day, he seemed thrilled and jumped up on the couch, was keen to explore the apartment, groomed his new furry friend, and both of them ate more hay.

However, the joy turned out to be short-lived for Håkon. After only 6 weeks as a happy rabbit, he died of an acute organ failure.

It is like stealing someone's life when keeping rabbits in the aforesaid manner, and we can only be happy that Håkon came to experience the love and enjoyment at least for a little while. In the end, life became more than just a comfortless, monotonous, lonely and boring existence. Even if it was only for a while, he got the chance to live and did not have to die alone in a tiny hutch.

Håkon and Sam. Photo courtesy of Linn Utseth-Hovelsen

Nancy and Trampe

Nancy was a single female who had lived as a house rabbit with her human family for 3 years. When they heard of Trampe, who needed a second chance, she was finally about to get a rabbit companion. Both of them were already neutered, so they were immediately brought to a garden neither of them had been in before. Both of them were released at the same time, but kept a safe distance before Nancy finally dared to move forward. Both rabbits seemed shocked, and they quickly ended up chasing each other. They were running, humping, resting and running for a while, and it all looked pretty normal for a bonding process. The caretakers had some cups of coffee, the rabbits relaxed and seemed to have come to an agreement, and eventually they packed up and drove home.

However, Nancy guarded her territory and was not as happy with an intruder in her house. As soon as he came close to her hiding spots and feeding station, she attacked him and gave him a nasty cut on his ear. The caretakers separated the rabbits instantly and went to the veterinarian. Trampe received the necessary treatment and both rabbits seemed happy to be together at the clinic. However, the joy did not last. At home Nancy once again attacked her new partner, so they were placed in pens next to each other until the next day.

During the night, the humans had switched litter trays and made sure the rabbits could stay close to each other without the possibility for physical confrontations. They had several meetings in a neutral room during the day, and were supervised all the time. Nancy was erratic and could not decide whether she should cuddle up with her new partner or attack him. The humans were worried, but I told them to endure. And it turned out that after a few days with careful supervision, moving back and forth between the separated pens and meeting room, the rabbits finally came to an agreement. Nancy and Trampe are today inseparable.

Females can be more territorial than males, and it is as a consequence sometimes slightly more challenging to move a male rabbit into a female's home than vice versa.

Combinations other than pairs

Trios, quads and larger groupings become progressively more complex, and it is impractical to discuss all the potential combinations here. As a general rule, whilst they can be stable groupings, with greater numbers comes greater opportunity for conflict, and the temporary removal of one animal, as for veterinary attention, may lead to disturbance and fighting within the group. All individuals in such groups must be neutered, as intra-gender aggression is much more likely in the presence of the opposite sex.

Greater space and environmental complexity is required to minimize these possible issues, and more feeding, watering and litter points need to be provided to avoid conflict. Hides or shelters require multiple exits, to avoid 'guarding' of them by dominant individuals.

Controversy exists as to whether two males plus one female is a more suitable grouping than two females and a male. It is likely that environmental factors and the individual nature of the animals are more significant than the genders.

> ### Challenging groups
>
> It can be difficult both to put together and to have a group of rabbits. If you do not have extensive knowledge, experience, space or resources, in addition to the possibility of keeping them in other arrangements if it is not working out, I generally recommend having a pair. A couple is far easier to house happily than a trio or a whole bunch of rabbits. Groups seem to be easier if they are growing up together. I know of several cases where one simply had to give up, despite excellent conditions and experienced owners.

Social Housing in Research

Many rabbits are utilized in laboratories, and it is of great significance to optimize the conditions for the individuals involved as far as possible.

Group housing is in this respect a significant enrichment, and will thus improve the rabbits' wellbeing. However, in laboratories this is mostly the case for females, since the hormonal behaviour of un-neutered adult males may lead to serious fighting, while does seem to be more tolerant of each other, even un-neutered.

Tension will occur, but the main reason why this form of cohabitation in these settings might work is probably that the arrangement is only for a period, not for a lifetime, and that the rabbits are being bred to be docile.

The likelihood for this to be successful is also increased by the use of sufficiently adapted pens, so that the rabbits are able to perform parts of their normal behaviour, which include gnawing, running, initiating and withdrawing from social and visual contact, and other play behaviour.

Groups must also remain stable to minimize conflict. It is usually not necessary to separate the rabbits, but if this has to be done they should be reunited with their group as soon as possible. If they need to be isolated, the rabbits involved should be able to maintain some contact based on smell and view, and later be reintroduced as normal.[3]

There is some controversy about the adult males involved in research, as they are primarily single housed. The benefits of living in groups or pairs rather than being single housed are vital for the species' wellbeing, and neutering of males and following social housing should be considered.

Stability

Bonded rabbits should not be separated. Their cohabitation is dependent on stability, and they communicate and orient themselves mainly by scent. Because of this, both rabbits in a pair should be brought together to the veterinarian if one of them has an appointment. Relocation and other changes in the environment may also cause temporary quarrels, caused by a confirmation of rank in the new area.

The knowledge of rabbits' social arrangements was first made known by R.M. Lockely in 1965.[4] His fieldwork and research, which was detailed in the book *The Private Life of the Rabbit*, was ground-breaking for understanding the species' social system. He observed a colony of rabbits, both above and below ground, and by using special glass-sided burrows he revealed a detailed complex social order.

Trio

Example 1

Melis was found on a roundabout outside a gym. An employee grabbed her and took her into the fitness centre. She was served dandelions, grass and water, and was kept company by various instructors until the local Protection of Animals group picked her up some hours later.

I decided to introduce her to my two male rabbits, so after neutering I moved her to a foster care where she would stay for some weeks. In order to improve the chances of a successful meeting I wanted her obvious hormonal behaviour to calm down prior to the bonding process.

I had arranged to stay in my sister's house and garden, and after 4 weeks I brought all three rabbits to this neutral area. I let them loose in the enclosure I had built in the garden, monitored the situation and was pleasantly surprised at how civilized they were. Harald was a bit angry and tense for about an hour, otherwise no offended rabbits' back-ends were observed.

We stayed in the house for 5 days and the rabbits were free-range in the kitchen and living room. Various boxes and hideouts were placed around, and they seemed to enjoy being a trio, even after the return home.

Example 2

Reidar was 6 years and about to be euthanized. He was healthy and neutered, and a nurse at the clinic took pity on him and brought him home. As she already shared her apartment with two female rabbits, Reidar could not get free access to the residence immediately and was accommodated in the bedroom. Ruth and Fiona were nevertheless immediately aware of the newly arrived male rabbit but did not appear to be significantly stressed.

The rabbits were slightly more dramatic than mine during the bonding. The process was conducted at the nurse's parents' home and thus on neutral ground. The animals chased each other for days, and especially Ruth was concerned about showing Reidar who was in charge.

Friction and quarrels between the two females also occurred during this period. This is likely to happen when the dynamics within the group are tested, whether in the form of a redistribution (individuals entering or leaving) of the group or if they are moved to another place. After the bonding, however, they were close friends again and had also become a trio.

Harald, Melis and Petter

One of his experiments highlighted the need for stability within the group. When he removed a dominant male and placed him outside his warren and territory for a while, the next strongest buck at once took over his position. The new leader defended his newly acquired property at full strength, which caused trouble if a former dominant rabbit returned.

This tendency may also lead to problems if one separates a bonded pair or a group of rabbits.

When a rabbit in a bonded pair disappears

When rabbits lose close friends they will mourn. Companions are very attached to each other and sudden solitude must be experienced intensely for rabbits, unable to influence their own situation. It is therefore highly recommended to offer the rabbit left behind a new friend as soon as possible. For the same reason, one should never separate a bonded pair.

> **Grief**
>
> When Håkon died Sam lost both his furry friend and his usual zest for life. He seemed puzzled and stopped eating, and his caretakers had to provide him nutrition through assisted feeding. He became sedentary and seemed apathetic, all visible signs of grief. His humans were surprised by his tremendous response, after all he had only been with Håkon for a few weeks, but after 10 days the house was still in mourning. Sam was not interested in any food and didn't move around.
>
> After 10 days, Sam's humans brought home a young female. He noticed the new rabbit immediately, became curious and alert, and jumped directly into the litter tray, where he began to eat his hay.
>
> A similar introduction as when he met Håkon took place, and the same night the two rabbits were eating together.

When former companions start fighting

Sometimes former companions begin to fight. Many owners are desperate, thinking that the rabbits must remain separated, whilst in most cases this can be resolved and the sudden enemies can become friends again.

There may be several reasons for such sudden clashes. Hormonally related behaviour can be said to be the main cause of rabbits' quarrels and disputes. Young rabbits growing up together and not being neutered in time will be affected by hormones, which causes exhausting cohabitation or possible dramatic fights. One must also consider whether the rabbit housing is sufficient to allow all the rabbits to live, jump, run, sleep and play there. Sudden changes in behaviour and hostility between former friends may also be a consequence of pain and discomfort. If problems persist in spite of recommended changes made by the owner, such as taking a car trip with the rabbits, rearranging the furniture and other steps carried out in the bonding process, a veterinarian should be contacted to examine the rabbits involved, looking particularly for signs of pain or hormonal abnormalities (e.g. adrenal gland disease).

> **Fighter's reunion**
>
> Even came to our home when he was a baby. The male and female in the house were sceptical at first, but after a bit of chasing they accepted him and eventually they became close friends. When Even was 10 weeks he began to show hormonal-related behaviour, he became more outgoing and apparently more annoying from Harald's point of view. He began to mark with droppings around the house, and we had him neutered when he was eleven and a half weeks old.
>
> They were a tight-knit trio who spent lots of time in proximity, even though they had a whole apartment at their disposal. However, after a year Harald probably had an accident with a wisp of hay and consequently an ulcer on one of his eyes. He was treated but still in pain, something that made him frustrated and more aggressive towards Even; he attacked him and bit his ear, and I had to separate the living room with fences and watch them the whole night. They behaved better the next morning, but Harald was still grumpy. It ended up with treatment of both rabbits at the clinic, a following car trip and reunion at a neutral place. The eye, ear and friendship recovered quickly after this.

Wild rabbit. Photo courtesy of Kevin Law, UK

Natural Behaviour and Social Life of the Wild Rabbit

Domestication has had little effect on animals' instincts. Even though rabbits in captivity will have little need to cooperate as far as eating and keeping guard, they still have an inherent need to live socially. Our conventional methods of raising rabbits have completely overlooked these needs, keeping them penned up alone in hutches. This type of confinement deprives the rabbit of social interaction, which in turn prevents them from displaying their normal behaviour.

Rabbits are gregarious and social animals. They belong to a group not only for survival but have a diverse and well-functioning cohabitation. The European wild rabbit, from which our domesticated rabbits originate, possesses the most complex and useful social relationship throughout the entire Lagomorph order. The order consists of hares, pikas and different rabbit species. Both hares and cottontails are species that can live by themselves; however, the European wild rabbit depends on a group to achieve safety and wellbeing.

In the wild rabbits live in colonies. Their underground tunnels can extend over large areas and be inhabited by hundreds of animals, while smaller colonies may settle in more concentrated burrows on a hillside. Regardless of colony size, they are subdivided into groups of from two to eight animals. Members in the respective groups are regulated, like the rest of the colony, in a linear ranking order. Rabbits have different positions in order of priority and have few conflicts once this is established. Dominant chasing and submissive retreat maintain the hierarchy, and this social arrangement provides an overall peaceful coexistence.

A group usually consists of more females than males. The does are usually tolerant of each other and spend much of their time in proximity with other females. However, usually friendly relationships can be disturbed when they are in heat and consequently argue over nest sites.

The females at the top of the ladder will have access to the most secure nesting burrows, and it is therefore important to earn an adequate position within the group. The social friction between females thus follows a hormonally related behaviour in order to ensure their offspring the best possible upbringing. These tendencies are also true for our domesticated rabbits, and pseudo-pregnancy may lead to serious arguments and hostility. Neutering eliminates this hormonal behaviour and is therefore recommended.

Does seem to be more territorial than bucks and can be very intolerant of trespass. Dominant females can be just as aggressive towards intruders or subordinates as dominant males. They can chase each other and fight; although it is rare that it turns out to be as violent as a fight between males can be.

Males are more stable compared to the females in terms of placement within the rank order but usually have more fierce battles than females in order to establish a hierarchy.

Threatening young males may be driven out by other males and must therefore attempt to enter new groups. The displaced rabbits obtain easier access to new established hierarchies in specific periods, outside of the breeding season. The groups are usually more relaxed and accommodating from August to January, while there is a more tension-filled atmosphere in the mating season that extends from January to August.

Observation of wild rabbits and studies of domesticated descendants shows that the species spends most of its time interacting with companions. Observation of free-ranging domestic rabbits revealed that the subjects spent 40% of their time in physical closeness with others,[5] and a study of laboratory group-housed rabbits has shown that they spend an average of 79% of their time in proximity with their companions;[6] however, they do not spend all of their time in direct visual contact with other rabbits,[7] so it is crucial to provide suitable living quarters.

Proximity and distance between colony members seems to be influenced by situation and time of the day, and females tend to spend more time with other females than males in general. However, they are all most tolerant of each other during resting, where they frequently have physical contact with companions. They also prefer sleeping alone sometimes, but as long as they have enough space and opportunities to choose where to lie down they regulate this peacefully. Frictions and quarrels within the group are mainly an issue when rabbits are kept together in inadequate confinement, e.g. hutches where they have no opportunity to retreat.

Rabbits also seem to be least tolerant of each other when they eat. In particular, when there are scarce resources, rabbits guard their own rations. This tendency can also be seen during feeding of pellets or when one of our house rabbits runs off with its mouth full of treats.

The Tradition of Keeping Rabbits Alone

Why is it then that many owners still keep their rabbits hutched up alone? This might be understandable knowing that the knowledge of rabbits, or rather lack of such, has been influenced by traditional rabbit breeding with an interest in fur, meat and exhibitions.

Commercial breeding systems feared the danger of infections when keeping rabbits in groups. They of course did not allow for successful social relationships anyway, with hutches and un-neutered animals, but explained the sole and hutched-up existence mainly on the basis of health.

In the case of infectious disorders, analysis repudiated this argument. Studies have proven that having rabbits in pairs or groups does not increase the chance of such infections. It is dirty environments and poor living conditions that lead to outbreaks of contagious diseases. Additionally, the diseases most common in group-housed rabbits are respiratory in nature. These are spread by droplet infection throughout a shared airspace, and individual cages offer no protection.

Rabbits who lived with their own species were also found to eat more than their single relatives, something that is not surprising since one often gets increased appetite in appropriate social company. However, they were not fatter, something that indicated increased activity.

Wild rabbits challenge each other. Photo courtesy of David Tunick, USA

Social conduct

Rabbits have a social conduct, which serves a number of different purposes. They can persuade other rabbits and gain social advantages such as grooming and cuddling, or increase their rank in the hierarchy.

Social aggression is an example of such tactics. The rabbits show dominance, obtain benefits and chase away possible threats by appearing to be strong and intimidating. This becomes especially important when changes occur in the group, such as relocation or when new members join the family. The 'boss' has to make sure that everyone knows their place by showing his authority, while the others must practise acceptable submissive behaviour.

This type of social aggression first becomes a problem in captivity when rabbits must live in abnormally small areas or when they are only allowed to meet occasionally. Without stability rabbits are unable to establish structures within the group. This leads to stress and permanent aggression between the members. With inadequate housing they have no means of getting out of each other's way, something which inevitably leads to aggressive confrontations. Therefore it is important to design the living quarters whilst keeping in mind the social needs of the rabbit.

In the wild, rabbits have a well-functioning social system. As a result of the ranking order, all rabbits know their place and fall into line with their duties and benefit from cooperation with the rest of the collective. They have an excellent ability to interact socially and a great need to stay together with other members of their species.

Karl Staff was found in a ditch. Between a pair of highways, he ran terrified back and forth on the impoverished lawn. He was skinny and in such poor condition that we did not dare to have him neutered until after a couple of months. Then he was sufficiently fit and physically almost like a new rabbit. However, he was sad and nervous and spent most of his time under the couch. He had access to an apartment but rarely ventured out and seemed scared.

After he was introduced to the somewhat more confident Ibsen, he immediately became a more exploratory and confident rabbit. It was love at first sight and he seemed relieved that he finally had a companion. He joined her on voyages of discovery in the apartment, groomed her coat as if he had never done anything else, ate more hay and became a rabbit who obviously felt safe.

Rabbit Cohabitation and Social Need

Living in a colony involves several advantages for a rabbits, such as increased defence against predators while grazing, cooperation and access to burrows, psychological benefits through mutual grooming, and the supply of potential partners and offspring. These behavioural interactions seem crucial for a rabbit's wellbeing and survival in the wild, but what about the domesticated rabbit living under confined conditions? Is a social life as important for a rabbit living in artificial and secure environments where the food is served? Is it not enough to be well fed and physically healthy?

A need is a requirement for something that is very important. Researchers distinguish between

Even and Melis often explore and play together.
Photo courtesy of Marit Emilie Buseth, Norway

ultimate and behavioural needs. The ultimate needs are those necessary for survival, such as food and water, while behavioural needs are essential for the individual's psychological wellbeing, such as being close to fellows of the same species. Animals have thus both ultimate and behavioural needs that must be met to maintain a satisfying life.

In the case of rabbits, knowledge of the species' social organization and studies on the behaviour of both confined rabbits and their free-ranging compatriots indicate that companionship and proper social contact is a behavioural need. Social contact is also found to be of equal importance to food.[8]

Adult female rabbits were used in an experiment[9] that examined whether the rabbits wanted to be close to each other or left alone. When given the opportunity, they all chose to make use of specially designed openings between two separate cages to be with other rabbits. Activity levels and other indications of welfare were raised relative to those who were not given this option and therefore remained alone.

But what about the rabbits ranking low in the social hierarchy? Do they prefer the company of others? Studies have proven that submissive individuals also prefer to have a companion,[10] even if they are being chased around sometimes.[11]

If you have a pair of rabbits it is thus important to be aware that both the submissive as well as the dominant have great joy and benefits from the cohesion.

The company of others seems to have an emotionally protective effect during stressful situations, and rabbits living in pairs or groups are shown to be less susceptible to stress than those living alone. Observation of solitary-housed rabbits revealed that they sat up more than social-housed rabbits, indicating increased fearfulness.[12] Abnormal behaviour or behavioural disorders, which typically are observed in single-caged rabbits, also seem to be absent in group-housed rabbits.[13]

It also appears that solitary housing in general reduces the rabbit's lifespan.[14]

The welfare of laboratory rabbits and their living conditions has been the subject of a relatively large

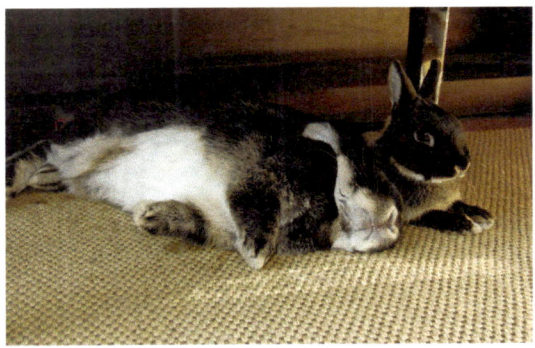

It is good to have a friend – Harald and Melis. Photo courtesy of Marit Emilie Buseth, Norway

number of studies for several years. Improvements of enclosures and enrichment in the form of hay have been evaluated. Well-equipped, large enclosures and chewing objects may counteract boredom to a certain extent, while the opportunities for companionship seem to be essential for the species' welfare.

Single-housed rabbits in traditional small cages indicated a higher degree of frustration in the form of stress-related behaviour, such as chewing on the bars, than rabbits living in groups. Single rabbits with space restrictions are also far more passive than the more playful and exploratory socially housed rabbits. This is easily transferable to our companion rabbits, as most people unfortunately know a depressed and apathetic rabbit in a hutch.

Rabbits that were so lucky to be able to live with other members of their species expressed more of the behaviour that indicates comfort, such as yawning, stretching and lying in positions that suggest confidence and wellbeing. They performed leaping and jumping more than their socially deprived relatives, who in turn showed a more depressive and quiet behaviour.

Read more about behavioural needs and welfare in Chapter 3.

Social Rabbits in Practice

Our domesticated rabbits have the same behavioural repertoire as their wild predecessors. A complex social behaviour allows rabbits to live and stay together. Further analysis also indicates that rabbits prefer to live in small social groups, both in captivity as well as in the wild. However, it looks like it is easier to make it work with two or three rabbits at home, rather than a larger group, so I recommend most people start with a pair. The more rabbits, the more space you need. A small groups' territory is on average 50 m^2, similar to a normal two-bedroom apartment.

Rabbits living together must have the opportunity to initiate as well as retreat from social contact. This means that they need sufficiently large areas with multiple sleeping places and territories in which to seek concealment. They also need places they can scent-mark by rubbing their chin on various objects. Odours are used in communication and help to maintain the hierarchy and one should therefore avoid sterilizing all nooks and corners while cleaning the rabbits' residential area.

Single Rabbits in Practice

It has long been common to keep single rabbits, but due to an increasing awareness and knowledge of rabbit welfare there are fortunately a growing number of people offering their rabbits appropriate

Mind your eyes!

Alfred, Edvard, Emma and Isa (dog). Photo courtesy of Hege Fjelde Tvedten, Norway

In a pet store, a sexually mature male rabbit was kept in a conventional cage. The staff might have thought it was a good idea to insert another young male rabbit into the cage, but that was sadly not the case. The hormonal and territorial rabbit attacked the newcomer, and the wounds were so severe that the employees thankfully separated the fighting rabbits.

The little one had severe ulcers on the eye, and it was decided that he should be euthanized. They just had to wait until they had time, so in the meantime he was placed in another cage.

After 3 days someone fortunately found him and rescued the injured little rabbit, and a few hours later the eye was surgically removed. A rabbit-friendly home was willing to care for him during recovery, and a prolonged treatment was started. The area without an eye had to be cleaned frequently, and rabbit Alfred had to be medicated and receive assisted feeding during the day and night.

Alfred stayed with his foster parents. The fur grew back again, and the little rabbit they had become so fond of was introduced to the house's two other rabbits. Edvard and Emma greeted the odd rabbit with one eye, and they soon became a happy trio. Alfred is a heroic rabbit whose experience illustrates the importance of neutering and how wrong things can go by simply placing a rabbit into another's cage. He is also a brilliant example of a rabbit that received a second chance and who lives a good life despite a disability.

The rabbits are free-range in an apartment, and Alfred has as good control as his companions with two eyes. That he should be euthanized due to foolishness is a frightening thought. He is a most charming and happy rabbit.

companionship and environment. Nevertheless, if for any reason it is not possible unable to keep them in a pair, it is important to know how to provide for the unaccompanied rabbit.

Single rabbits should always be allowed to live indoors with the rest of the family. A daily visit when you go out with food or to snuggle with your companion for half an hour is not enough to cover the animal's social needs. However, they can still enjoy life if they have very dedicated humans who spend a lot of time with them. A confident free-range house rabbit can wander around, seek out humans and benefit from their company. Most people are aware that dogs are dependent on a social life. One often goes home to the dog after work or school, and many think accordingly that the cat should not be home alone too much either. The same should apply for a rabbit.

Rabbits living in sufficiently devoted and attentive homes are doing fine. However, there is no doubt that they will do even better with another representative of the species in the house. Neither humans nor other animal species can replace another rabbit.

The species is dependent on another conspecific to communicate and express a full range of social behaviours. In addition, humans are not generally available for social interaction at the same time of day as rabbits, even if they are at home all the time, due to the their diurnal rhythms and need for being active at dusk.

However, once you have had two rabbits and seen how much joy they have with each other, I promise that you will never let a rabbit live alone ever again.

Aksel and Trulte. Photo courtesy of Marit Emilie Buseth

Rabbits and Babies

Many wave their rabbits goodbye when their daily life changes. Changes may involve moving house, changing jobs, getting a dog or having a child. When acquiring animals one should include them in future plans and possible changes later in life, and it is also usually easy to incorporate the rabbit in a new situation, as when having a baby.

Petter was an only child and a single rabbit for nearly 6 years. On a somewhat poor foundation I had previously let him meet a few other rabbits with little success. He was, however, a remarkably harmonious and well-adjusted rabbit that I spent lots of my time with. He jumped up on the desk when I was working, he sat on the sofa when I watched TV, he lay in my bed when it was time for me to go to sleep, and he inspected all my guests who came to visit.

Petter was probably a content rabbit, but I have no doubt that he became even happier with fellow members of his species. Despite my efforts, I was in fact never a rabbit. I could not give him the same as Harald apparently did when he moved in.

Harald was just a baby when he was rescued, and Petter found a new best friend. He was suddenly transformed into a proper rabbit. He was like a youth again, he was more active and popped around the house, wriggled more and seemed enthusiastic about his furry friend, who incessantly licked and groomed his coat.

As many others, I had been afraid that Petter was going to change and forget about me, but we retained our close relationship as long as he lived, even with eventually three rabbits in the household.

Social Rabbits

Mio and Tobias and a practical litter tray and eating station. Photo courtesy of Marit Emilie Buseth, Norway

Human baby Tobias was born while My was a single rabbit, and the new parents were initially uncertain about how it would turn out with both a baby and a rabbit in the small apartment. They worried about whether they were able to take care of the rabbit when baby bath, toys, diapers and cries filled the house.

Tobias was like most children and required a degree of care and attention. My, on the other hand, was still happy to sneak around as she pleased, nibble hay when it suited her and enjoyed being free-range. She led to no more work after Tobias came into the house and seemed indeed as if she liked the little boy. She was babysitting when his parents went into the kitchen, and Tobias could gurgle and laugh in his tilt chair, while a faithful rabbit sat beside.

Since everything went so well, the family decided to help another homeless rabbit as well, and thus it was that Mio moved into the house when Tobias was 7 months. The former litter tray was replaced with a more child-safe system. By moving the litter tray and hay dispenser into a specially adjusted basket with a suitable door, they prevented Tobias from eating droppings and pulling hay around the living room, while the rabbits were given a secret and quiet place where they could eat and rest without being disturbed.

For Tobias it became completely normal to live with two rabbits. He developed an understanding and respect for the two animals that plodded around the house, and both he and his parents took care of them. He learned that he must pay attention to the animals, that he could not pick them up or carry or chase them around, and the rabbits also learned that the boy was not a threat. They could go away if they wanted. Tobias also showed a great pleasure and interest in his two furry friends.

Adaptation of the area can thus be all it takes to continue to live with animal friends. A new arrangement of the litter tray may be all it takes to combine free-range rabbits and toddlers.

There is no health risk associated with being pregnant or keeping babies and rabbits together. Rabbits rarely carry bacteria pathogenic to humans, and their droppings and their food, even if eaten, are unlikely to cause harm. Some studies also suggest that children raised around animals are less prone to allergies.[15]

Rabbits with Other Animals

Guinea pigs

Many rabbits and guinea pigs live together and are apparently good friends. However, there are several reasons not to house these species together. First of all, a guinea pig and rabbit will not be able to communicate appropriately. They may snuggle together, but they will not understand each other's

Gaston and Dinka having a siesta. Photo courtesy of Katarina Vallbo, Sweden

scent signals or body language, such as position of the ears, that can express discomfort, dissatisfaction, dominance or other signals that it is important for animals to convey. Guinea pigs are also extremely social individuals and should live with fellow members of the species.

Rabbits can harass and torment guinea pigs, even without us understanding what is going on. Although the two species look equally cute and harmless, the guinea pig has no defence against the rabbit. Many guinea pigs also hide injuries caused by rabbits, either in the form of a bite or kick with the rabbit's powerful hind legs.

Rabbits are often bearers of *Bordetella bronchiseptica*, a bacterium that does not cause any harm for the rabbit but that often develops into a serious and fatal disease in guinea pigs. Guinea pigs should not be exposed to rabbits as a potential source of infection.

Guinea pigs and rabbits also have different nutritional requirements. Guinea pigs have a higher protein requirement than rabbits, and since they are unable to synthesize Vitamin C, they need a dietary source and supplement.

If nevertheless you have rabbits living with cavies, it is important to facilitate their living area

Molly and Ture. Photo courtesy of Emma Almquist, Sweden

properly. The guinea pig should have a hide to which the rabbit does not have access, for example by making the entrance very small. This will provide the cavy with an opportunity to escape a possible dominant and quarrelsome rabbit. It will also be possible to serve the guinea pig its food and Vitamin C supplement in this hiding spot.

Guinea pigs may bully rabbits as well. It is not uncommon for guinea pigs to sit on rabbits and barber the fur of the head and back. This is more difficult to prevent, but providing higher levels that the rabbit may climb/jump on to, but the guinea pig cannot, may help provide a 'safe' area.

Dogs and cats

There are many examples of dogs or cats that have become friends with a rabbit. It is especially common for cats and rabbits to share a household. However, it is impossible to give any advice that ensures the welfare and lives of rabbits if these species are going to live together, as it depends on each individual and the particular circumstances.

One must first of all be aware that the rabbit is a prey animal that is careful and naturally afraid of predators that could pose a threat, and that both cats and dogs are carnivores with their hunting instincts intact.

Mia was fortunate and had a nice apartment and kind adoptive parents. She had not had a good start in life, but was rescued and given a second chance when she was 7 months old. Her new owners were attentive and loving, gave her free access to the house, were patient and tried to give her a sense of security, and she became a more confident rabbit. She ran around on the carpet, was very fond of her castle and the new baby that eventually came into the house. When Tobias was lying on the floor, Mia often sat nearby, almost like she was babysitting.

But she was still too nervous to let people provide her with cuddles. She always shied away when they gently tried to pat her nose or otherwise wanted to be close to her. She rarely dared to go around and explore the room, and her humans asked me what more they could do for her to be less afraid and benefit her socially, and then Mio moved into the house.

When we decided they should meet for the first time, we placed both the neutered rabbits in their carriers, drove half an hour and let them be introduced at a friend's house. It was a neutral location for both of them, and after they had ventured out of the cage they saw ... a rabbit! Stunned and inquisitive they sniffed around, and after a few seconds Mio was heading away to greet Mia. Hopefully he laid his head on the ground in front of her, while she quickly took off. Confused and frightened, she ran away with a purposeful rabbit behind.

They ran around chairs and tables, jumped on carpets, in and out of their carriers, up on the sofa and around in circles. We sat on the floor and watched. The female took off to escape from the eager animal, which reached her and started to hump. She jumped forward with an determined Mio behind. They rampaged around for about an hour, only interrupted by mutual ceasefires and breathing breaks, before we put them in a joint carrier and drove home.

They were sitting stock still next to each other, apparently unaffected by the fellow passenger. They actively ignored each other after returning to their home as well, and Mia was busy with overlooking the newcomer who suddenly sat in her litter tray eating hay.

After an hour with disregard, however, she went over to say welcome and licked his head.

Mio. Photo courtesy of Marit Emilie Buseth, Norway

One should never let a dog or cat greet a rabbit sitting in a cage. Rabbits must have the opportunity to get away from any threat, and being hutched up represents a powerful stress factor when a predator is outside. The rabbit is fully aware that it cannot escape and might be terribly frightened.

If introducing a dog and rabbit, one should therefore allow the rabbit to move freely in the room and keep the dog on a leash. If the rabbit is familiar with the environment, it will know where to run to arrive safely in a hiding spot, and the dog can easily be controlled when it is kept on a leash.

Train your dog in advance, to ensure that he follows your 'Leave it alone' command. If he looks calm and the animals are either positively interested or ignore each other, one can try to have them together without harnesses, but always under supervision and a sample of rewards in your pocket.

If introducing a cat and rabbit, it is also important to allow the rabbit to be free-range and with the ability to hide if necessary. Although your own house-cat or family dog may be friendly towards your rabbits, you need to be aware that strangers' cats and dogs may be able to attack and kill rabbits if left outdoors unattended.

However, there are many stories about rabbits that take over control and chase cats around the house, and some rabbits clearly think they decide, but this is obviously due to the rabbit's high self-esteem, as they naturally enough have nothing to threaten the small tigers.

One can never ensure the predator's hunting instinct will not reawaken, and one should therefore warn against keeping rabbits with dogs and cats unsupervised.

Ferrets

Ferrets are small carnivores. In the wild, their wild counterparts, stoats, weasels and ferrets, will eat small rodents, frogs, birds and eggs, but domesticated ferrets have also been used in rabbit hunting. Rabbits will never feel safe and secure close to such an efficient hunter, and all contact between the two species should be avoided.

Rabbits appear to be far more fearful of ferrets than cats and dogs, which should also be taken into account when visiting a veterinarian.

My rabbits are mostly safe and satisfied when visiting the veterinarian. They are familiar with the clinic and walk around exploring when we are there. However, once when we were there they started thumping, sweeping around in panic and hid under a cabinet. It turned out that a ferret was in the examination room next door.

Notes

[1] Held, S.D.E., Turner, R.J. and Wootton, R.J. (1995) Choices of laboratory rabbits for individual or group-housing. *Applied Animal Behaviour Science* 46(1), 81–91.

[2] Hull, W.L., Brooks, D.L. and Bean-Knutdsen, D. (1991) Response of adult New Zealand white rabbits to enrichment objects and paired housing. *Laboratory Animal Science* 41(6), 609–612.

[3] Hawkins, P., Hubrecht, R., Buckwell, A., Cubitt, S., Howard, B., Jackson, A. and Poirier, G.M. Refining rabbit care. A resource for those working with rabbits in research. Available at: http://www.rspca.org.uk/servlet/BlobServer?blobtable=RSPCABlob&blobcol=urlblob&blobkey=id&blobwhere=1213709292078&blobheader=application/pdf (accessed 10 October 2011).

[4] Lockley, R.M. (1965) *The Private Life of the Rabbit*. A. Wheaton & Co, Exeter, UK.

[5] Vastrade, F.M. (1987) Spacing behaviour of free-ranging domestic rabbits, *Oryctolagus cuniculus*. *Applied Animal Behaviour Science* 18(2), 185–195.

[6] Gunn, D. and Morton, D.B. (1993) The behavior of single-caged and group-housed laboratory rabbits. *Proceedings of the Fifth Federation of European Laboratory Animal Science Association (FELASA) Symposium*, 80–84. Cited in: Boers, K. et al. Comfortable quarters for rabbits in research institutions. Available at: http://www.awionline.org/pubs/cq02/Cq-rabbits.html (accessed 10 April 2013).

[7] Seaman, S. (2002) Laboratory rabbit housing: an investigation of the social and physical environment. Available at: http://www.ufaw.org.uk/pdf/phhsc-schol1-summary.pdf (accessed 15 February 2013).

[8] Seaman, S. (2002) Laboratory rabbit housing: an investigation of the social and physical environment. Available at: http://www.ufaw.org.uk/pdf/phhsc-schol1-summary.pdf (accessed 15 February 2013).

[9] Held, S.D.E., Turner, R.J. and Wootton, R.J. (1995) Choices of laboratory rabbits for individual or group-housing. *Applied Animal Behaviour Science* 46(1), 81–91.

[10] Held, S.D.E. (1996) Group-Housing of Female Laboratory Rabbits – Studies on Behaviour and Immunocompetence. PhD dissertation, University of Wales, Aberystwyth, UK. Cited in: Boers, K., Gray, G., Love, J., Mahmutovic, Z., McCormick, S., Turcotte, N. and Zhang, Y. (2002) Comfortable quarters for rabbits in research institutions. Available at: http://www.awionline.org/pubs/cq02/Cq-rabbits.html (accessed 10 October 2011).

[11] Batchelor, G.R. (1999) The laboratory rabbit. In: Poole, T. and English, P. (eds) *The UFAW Handbook on the Care and Management of Laboratory Animals*, 7th edn. Blackwell Science, Oxford, UK, pp. 395–408. Cited in: Boers, K., Gray, G., Love, J., Mahmutovic, Z., McCormick, S., Turcotte, N. and Zhang, Y. (2002) Comfortable quarters for rabbits in research institutions. Available at: http://www.awionline.org/pubs/cq02/Cq-rabbits.html (accessed 10 October 2011).

[12] Schepers, F., Koene, P. and Beerda, B. (2009) Welfare assessment in pet rabbits. *Animal Welfare* 18(4), 477–495.

[13] Loeffler, K., Drescher, B. and Schulze, G. (1991) Einfluß unterschiedlicher Haltunsverfahren auf das Verhalten von Versuchs- und Fleischkaninchen. *Tierärztliche Umschau* 46, 471–478. Cited in: Boers, K., Gray, G., Love, J., Mahmutovic, Z., McCormick, S., Turcotte, N. and Zhang, Y. (2002) Comfortable quarters for rabbits in research institutions. Available at: http://www.awionline.org/pubs/cq02/Cq-rabbits.html (accessed 10 October 2011).

[14] Schepers, F., Koene, P. and Beerda, B. (2009) Welfare assessment in pet rabbits. *Animal Welfare* 18(4), 477–495.

[15] Warner, J. (2011) Pets may reduce children's allergy risk. Children who had a dog or cat as infants less likely to become allergic. Available at: http://www.webmd.com/allergies/news/20110613/pets-may-reduce-childrens-allergy-risk (accessed 10 March 2013).

Gaston (cat), Dinka and Gabriella.
Photo courtesy of Katarina Vallbo, Sweden

Kalle from snout to tail. Photo courtesy of Axel Hunstad, Norway

5

From Snout to Tail

Rabbits are not always easy to understand. They don't bark and may not show signs of pain or discomfort as clearly as a cat. Being a prey animal, a rabbit will always do its best to hide the fact that it is in pain or injured and is thus potentially vulnerable. With enemies in the air, beneath the ground and lurking behind every bush, the rabbit will try to minimize its chances of being seen and eaten. It will not want to attract unnecessary attention by revealing itself as being weak and therefore the easiest target in the group. This is also why your house rabbit will pretend that it is strong and healthy, even if it has been in pain over a period of time. When the suffering has become so severe that the rabbit is no longer able to hide that it is in pain, this means that the condition is so advanced that immediate treatment is necessary.

Rabbits are subtle animals. It is therefore unfortunate that children have the responsibility of caring for rabbits, as they in general lack the ability to detect early signs of illness and discomfort, and thus overlook important signs that eventually lead to more serious injury.

Anyone who lives with a rabbit must know how to identify the symptoms that something is wrong, know what it might mean and what to do, which in most cases means a trip to the vet.

Rabbit Anatomy

Skin, fur, paws and tail

Wild rabbits have a short-haired coat that is easy to groom and keep clean. Domesticated rabbits, however, have varying fur attributes, as some breeds have longer, softer or thicker hair than others. Regardless of the exterior, all rabbits have a thin and sensitive skin and a soft undercoat that is covered with a dense and protective fur. They have thin skin with no calluses on the soles of the feet,

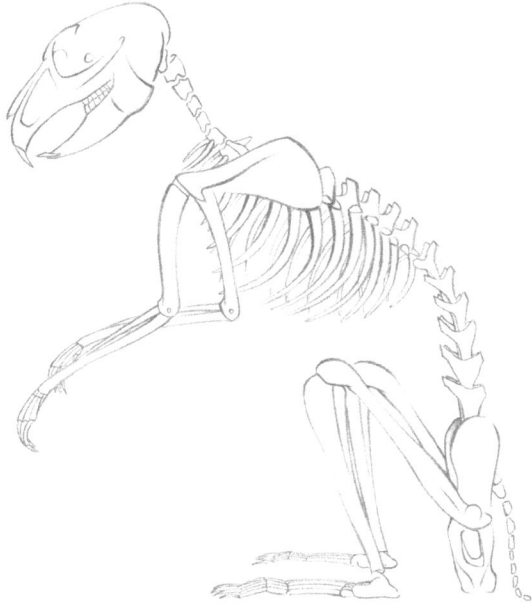

Illustration courtesy of Nils Erik Werenskiold, Norway

and rex breeds are particularly susceptible to sore hocks and require care with selection of substrate and regular checking of the feet, due to their very short, almost velvety fur. Breeds with excessive fur, such as cashmere, angora and even the more popular lionheads, are high maintenance breeds, requiring regular grooming assistance. There are many long-haired rabbits suffering from lack of care. Painful felted wool and hair must often be shaved away when rabbits come in to rescue centres. Individuals with such a long coat are not suitable to life outdoors in damp areas or in a cage with shavings sticking to the fur.

Long fur around the rabbit's face prevents visibility and consequently makes the rabbit more

© M.E. Buseth and R.A. Saunders 2015. *Rabbit Behaviour, Health and Care*
(M.E. Buseth and R.A. Saunders)

nervous and jumpy, and a haircut is recommended for better vision. Be careful not to cut off their whiskers, as these are required for sense perception around the head, for moving through small spaces and feeding.

Rabbits normally shed twice a year, and the process starts at the head and spreads down the body towards the tail. The dramatic hair loss can appear to inexperienced eyes as a sign of disease, whereas it is the rabbit's natural change of coat. It is important to help the rabbit remove excess fur from the body during moulting periods. Most rabbits will benefit from daily grooming, and longer haired rabbits may require trimming as mentioned above.

Rabbits do not have footpads like cats and dogs. Instead, the feet are covered with thick fur that provides a spongy shock absorber between skin and ground. With such vulnerable paws, it is important that one never removes the fur on the underside of the rabbit's feet. The foot anatomy also makes the rabbit vulnerable to damage if living on wire floors or other rough surfaces. Wire floors should be avoided.

> Rabbits should never have wire-mesh flooring in their cage. The foot anatomy makes the rabbit vulnerable to damage if living on wire floors, and they can suffer from sore hocks and broken claws. It also prevents normal movement.

Wild rabbits burrow into underground passages to hide from predators, protect their kits and to shield themselves from extremes of cold or heat. To be effective diggers, they have fast-growing and sharp claws. Rabbits have five toes with claws on their front paws, although the medial (or inside one) is small, and four on the rear paws.

The rabbit has loose skin on the neck, which forms a dewlap on the underside of the neck in females, particularly those of the meat breeds. This is most evident in un-neutered females. An obese rabbit, even a male, will develop an unhealthy and troublesome dewlap, and once they develop they can be difficult to lose by dieting. In some cases, surgical removal of such folds may be necessary. The same rabbits also have extra folds of skin beside the genitals, making them particularly vulnerable to urine scalding.

It is worth mentioning that rabbits do not have sweat glands in the skin, other than a few on the lips, which means that they cannot sweat. They also cannot pant, as they breathe through the nostrils. They are therefore at risk of heatstroke at high temperatures and in direct sunlight.

The rabbit has a small tip of tail covered with fur. Rabbits can move and control the tail and it is important for communication with other rabbits. Scent glands located under the chin, on either side of the genitals and the anus also play the main role in the rabbit's scent-based communications.

Muscles and skeleton

Molly is a fit rabbit. Photo courtesy of Emma Almquist, Sweden

Rabbits have a very light and fragile bone structure, and the skeleton forms only 7–8% of body weight. The species' brittle, almost bird-like skeleton makes them vulnerable to caging conditions with little or no opportunity for normal activities, such as running, hopping or even just sitting up on their hind legs. Rabbits housed in standard cages will lack proper motor coordination, and will be prone to muscle wasting and weakness in the bone structure. In rabbits that have developed such osteoporosis, the brittle bones are more at risk of fracture and damage, especially those of the lower spine.[1] Since the rabbit is excellent at hiding pain, it is difficult to know whether your companion rabbit actually has a breakage or other abnormalities in the skeleton, and the best one can do is to prevent such problems by providing a suitable living environment and diet. The species' anatomy implies their capacity and need for exercise, and it is crucial to offer them suitable living arrangements.

The spine is naturally curved; the front legs are short, while the hind legs are extremely powerful.

Hans. Photo courtesy of Aksel Hunstad, Norway

The rabbit's skull is large compared to the body. The shape of the head varies according to the breed, but naturally rabbits have long faces and noses (doliocephalic) with plenty of room for their dentition. Certain breeds are to a greater extent prone to dental problems due to their short faces (brachiocephalic) and it is important to prevent this by means of a diet that is based primarily on hay.

Ears

The ears are the rabbit's thermostats. The species is not able to sweat. Damaged ears, such as those with missing tissue, from injuries early in life, cutaneous myxomatosis, or problems associated with the use of their veins, can thus cause problems in getting rid of excessive heat or adjusting heat loss.

Lops are more prone to ear infections than breeds with more natural upstanding ears.

Eyes

Rabbits have large eyes, which are located on opposite sides of the head. It is an adaptation among prey animals whose survival depends on successfully keeping track of predators both above and below ground, and in the air. The eyes' position enables the rabbit to have an expanded field of view and to monitor a near 360° angle. Rabbits should therefore be able to look both backward and to the side, and it is unfortunate that this is not so easy with long hair and huge hanging ears.

Rabbits thus see both backward and to the side, but have a blind point straight in front of the nose and mouth. This blind spot means that the rabbit is dependent on their good sense of smell and whiskers around the mouth to find their food.

Rabbits have a third eyelid that can be drawn up to protect the eye. One can sometimes see it in the corner of the eye, but if it is visible at all times it

From Snout to Tail

might press on the eye caused by inflammation or problems with the tooth roots, or an eye infection. Severe gastrointestinal disorders may also make the third eyelid visible.

Nose and respiratory tract

Most people have seen rabbits sniffing persistently. They have very sensitive nostrils that twitch at a rate of 2–120 times per minute, depending on whether they have a lot to keep track of or not. They take 30–60 breaths per minute, although this commonly rises rapidly in rabbits undergoing stress.

Rabbits are obligate nasal breathers and cannot breathe well through their mouth. If a rabbit is gasping for air, something is seriously wrong and a veterinarian must be consulted immediately. This usually indicates a nasal cavity blocked with pus, blood or a tumour, and is generally a bad prognosis for the rabbit.

Rabbits orient themselves mostly by means of sense of smell and surrounding whiskers, so they do not like to be touched on the nose, which also must be dry and clean. The lungs are relatively small and thus vulnerable to complications from pneumonia and other lung diseases. Draughts must be avoided.

Dentition

Rabbits have an advanced and highly specialized dentition consisting of four incisors in the upper jaw and two in the lower jaw. In addition to the front teeth rabbits have several cheek teeth: three premolars and three molars on each side of the upper jaw, and three premolars and two molars on each side of the lower jaw. There are 28 teeth in an adult rabbit's mouth in total. These have open roots, i.e. soft tissue that forms new dental cells so that the teeth grow throughout their lives.

Acquired dental problems are often related to incorrect feeding. Rabbits are obligate high-fibre herbivores and must eat hay or grass to maintain a healthy intestinal function and a viable dentition.

Most of the plants in the grass family have microscopic silicates in their leaves, which are highly abrasive, wearing down the teeth in the animal eating them. In addition, the grasses' shape and plant fibre content (cellulose and the digestible fibre) requires a figure of eight grinding jaw movement for abrading the plant material to a consistency ready to be ingested, with both side and back and forth movements in the lower jaw relative to the upper jaw. This wear pattern is instrumental in maintaining the shape and occlusion of the teeth and avoiding the development of acquired dental disease. Eating insufficient amounts of grass and hay is a major factor in development of malocclusion, which in turn can lead to serious and painful dental diseases.

Read more about dentition in Chapter 6.

Sandor and Ru are healthy rabbits eating hay. Photo courtesy of Katarina Vallbo, Sweden

Gastrointestinal tract

The rabbit's stomach and intestinal tract is designed to digest and utilize nutrient-poor and fibre-rich plant foods. The species' intestine is divided into several specialized departments, and there is in addition a very special community of microorganisms (bacteria, fungi and single-celled animals) that digest plant fibre and convert it to volatile fatty acids that rabbits then use to build their own body cells. This type of digestion may be called hindgut fermentation and digestion, and is quite similar to the way horses utilize plant foods.

Rabbits also have a specialized caecum and colon. The rabbit passes out both normal faeces, almost entirely made up of indigestible fibre, and caecal content pellets known as caecotrophs, which must be re-ingested to complete their digestion and absorption. These caecotrophs are rich in vitamins and protein, and rabbits eat them to utilize their meal. This process is called coprophagy and is the reason that the rabbit can survive on a nutrient-poor diet.

The digestive system and diet is thoroughly discussed in Chapter 6.

The cardiovascular system

The heart is relatively small and constitutes only 0.3% of the total body weight. A consequently relatively lower cardiac output makes the rabbit unable to run the same distances as a hare, which, due to a larger heart, gets more oxygen and energy pumped out to the muscles.

When taking blood samples or inserting catheters, the blood vessels and veins are most easily accessible on the ears and hind legs.

The normal heart rate is 130–350 beats per minute. This will increase due to pain, stress or high temperature, and also depends on the size of the rabbit, with larger breeds having slower heart rates and smaller breeds even faster than this.

The urinary tract

Being herbivorous leads the rabbit to excrete alkaline urine. The species' normal urine pH is 8–8.2[2] as opposed to the acid urine of carnivorous animals.

In addition, rabbits have a special method for controlling the calcium level in the body. Whereas other animals only take up the amount they need, the rabbit will absorb almost all the calcium available in the food they eat. Then they excrete the excess to the urine via the kidneys. This makes the rabbit urine rich in calcium and the animals prone to develop kidney stones and large amounts of calcium rich sediment when the calcium settles out.

The variable urine colour is a result of porphyrins, groups of pigments or colouring substances in the plants that the rabbits are eating. The urine appearance normally varies from creamy white to bright yellow or red. Very creamy, white urine indicates a diet too rich in calcium.

Sexing and genitalia

Immature rabbits can be difficult to sex. Wriggling individuals of 4–6 weeks of age, with few obvious signs to distinguish between males and females, can make it challenging to determine gender at an early age, and thus many does have male names and vice versa. As rabbits unfortunately often are sold at such a young age as 6 weeks, it is important to be aware that the given gender is not always correct. Consult a veterinarian or a rabbit-experienced person to determine the sex properly before males are sexually mature at 10 weeks of age, so fighting and pregnancy is prevented. A moderately uncommon condition, known as 'hypospadia', is present in some strains of mini and dwarf lop rabbits in the UK. In this, the opening of the penis appears more slit-like, and therefore resembles a vulva. It is very easy to mistake a male with this condition for a female, until the testes are present.

Adults are usually straightforward to distinguish, since the testicles descend at 10–12 weeks, although they can be retracted into the abdomen during stress or disease.

It is hard to teach how to decide gender by the use of images and text, and the penis in the male and the vulva in the female will for a novice look quite similar. The vulva is less round and a bit shorter than the penis, but it is difficult to see without comparison.

Both sexes have deep skin folds on either side of the genitals and the rectum. These are scent glands that emit a distinctive odour that is an important part of species communication.

It is possible to clean these folds to remove the waxy, malodorous exudate, but is not normally required. If necessary, use a moistened cotton swab to remove the dark musk. Be aware of the thin and fragile skin and be cautious.

Before You Attend a Clinic

It is extremely important that the veterinarian treating the family pet has extensive knowledge and experience with rabbits. One cannot derive an understanding of rabbits based on knowledge about dogs, cats and cattle. A rabbit-competent vet should also know what is normal and abnormal rabbit behaviour to ensure the best possible health and welfare. They must be able to detect the subtle signals that indicate disease or dissatisfaction, as well as be able to communicate this to the owners so that they can be better qualified to take care of their companion animal.

This knowledge about rabbits is available in veterinary textbooks, journals and CPD training. Veterinary clinics that purchase the necessary equipment for improved care of rabbits and provide further education to their employees should be recognized.

It is important to find a qualified veterinarian with rabbit experience and knowledge before the animal is sick, so you know where to call when an acute situation arises. Enquire amongst friends and local sources of information about different vets in the area regarding their knowledge of exotic prey species, and contact them to ask specific questions.

You may wish to ask what experience a vet has in treating rabbits, whether they have any further qualifications in either 'exotic' animals or in the specific discipline with which your rabbit has a problem.

In addition to annual vaccination it is recommended to bring apparently healthy rabbits for annual health checks, and to consult the veterinarian as soon as one discovers something wrong. Bringing rabbits in before something goes wrong is a useful way to develop a relationship with your veterinary practice, and familiarize yourself with their hours of business, location, policies, etc.

Obtaining insurance to cover veterinary bills is something to consider seriously. Look at the range of policies available in your country, compare the relative prices, but also look at the level of cover provided and the list of potentially excluded conditions before making a decision.

Travelling to the Clinic

As with all transportation, it is important to minimize stress relating to the trip. Rabbits should be transported in a carrier that is large enough so that the rabbit can turn around, but not so big that it will be thrown from side to side during the journey. It may be secured within the vehicle with a seatbelt or material such as coats or pillows. One must always make sure that dry hay is available, and on longer journeys water must also be available. Rabbits are herbivores and will instantly feel unsafe if they are placed in a bare travel box. They are dependent on nibbling hay and even know that they will not survive long without this opportunity. For the rabbit to

---

Details of what courses and training your vet has attended are useful. Regular attendance at rabbit or exotic courses is a positive attribute. Membership of specific organizations, such as the RWAF (Rabbit Welfare Association & Fund) and the BVZS (British Veterinary Zoological Society), shows a willingness to learn and become further involved in such species.

Consider also what happens when your usual vet is on holiday. Do others in the practice share their interest in rabbits? And also ask what happens in 'out of hours' situations: emergencies at nights and weekends? Will your usual vet see your rabbit or will you be advised to travel to an out-of-hours provider? And if so, how rabbit friendly are they?

Your vet may already have a considerable library of textbooks, journals and other sources of information to consult, but this is a useful list, which you, of course, may also wish to purchase:

James Carpenter (2012) *Exotic Animal Formulary* (4th edn)
Frances Harcourt-Brown (2002) *Textbook of Rabbit Medicine* (1st edn)
John Chitty and Frances Harcourt-Brown (2013) *BSAVA Manual of Rabbit Surgery, Dentistry and Imaging*
Anna Meredith and Paul Flecknell (2006) *BSAVA Manual of Rabbit Medicine and Surgery* (2nd edn)
Anna Meredith and Brigitte Lord (2013) *BSAVA Manual of Rabbit Medicine*
Katherine Quesenberry and James Carpenter (2012) *Ferrets, Rabbits and Rodents* (3rd edn)
Richard Saunders and Ron Rees Davies (2006) *Notes on Rabbit Internal Medicine*
Molly Varga (2014) *Textbook of Rabbit Medicine* (2nd edn)

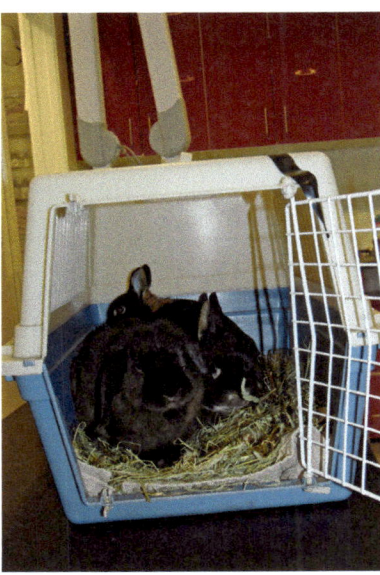

Everyone is going. Photo courtesy of Marit Emilie Buseth, Norway

feel safe it is thus essential to have sufficient food in the carrier.

Rabbits will often urinate in the cage due to stress. Make sure you supply a good surface, such as an absorbent towel or incontinence sheeting , so the rabbit does not get urine in the coat.

In addition to feeling relatively well hidden, they should be able to see out. One should also try to avoid inquisitive humans or animals sniffing at the carrier or trying to touch the rabbit, especially if it is sick. If you are in noisy and crowded environments with scary smells and pestering dogs, you may want to put a blanket over part of the travel cage.

If your rabbit is with one or more rabbits, all within the group should be taken to the vet. This is to ensure that the rabbits do not quarrel when coming home, which might occur if one rabbit has been away and has been covered with unfamiliar scents at the clinic. The rabbits will also seek solace in each other, both on the road and at the vet.

When reaching the clinic, one should emphasize that the rabbit should not stay in the room where dogs, cats or ferrets are present. The rabbit's predators should not be allowed to make the prey more scared and stressed. If your vet has no more rooms available, it is better to put the carrier in a broom closet or stay in the car rather than throwing them to the wolves.

A Routine Check-up

At a routine check-up the veterinarian will examine the overall health of the patient. They will go over the rabbit's body, from snout to tail, in order to assess whether the animal is healthy or has symptoms of illness. It is important for the vet to remember that the rabbit, to an even greater extent than normally, will try to camouflage the pain and disability as it is particularly watchful outside its familiar surroundings. It is not uncommon to see rabbits 'recover' when you get to the doctor. A veterinarian, who has experience with prey animals, will be familiar with this mechanism and know about the way rabbits act.

The veterinarian will observe and look for signs of whether something is wrong with the rabbit, and will examine and feel the body and make sure to obtain the rabbit's accurate weight. In case medications or anaesthesia become necessary, one must have knowledge of the weight so that the dosage is adjusted correctly. When the rabbit is handled and examined, it is important that you as the owner step aside so the rabbit does not get further stressed. Rabbits can feel that you are upset and be affected by this. Let the qualified vet take control and listen to what they have to say. Whilst it can be upsetting to step away and let a member of staff restrain your pet, many rabbits will attempt to jump towards their owners, to apparent safety, and can injure themselves so doing.

The veterinarian will first talk to you about the rabbit. If you suspect that something is wrong, it is important to be clear on what you think is different and how long you have noticed this change. They will also want to know something about the rabbit's daily life, how the rabbit is kept, what it eats and what you offer in the way of activities. It is often helpful to write down notes to refer to in the consultation, to avoid forgetting details, and some vets may provide checklists or history sheets for you to fill in before an examination. A number of rabbit ailments are a direct or indirect consequence of feeding and housing, and a vet should be familiar with the rabbit's living conditions in order to provide as good a treatment as possible.

The vet may place the rabbit on the ground to watch it walk and hop around. As well as information about mobility, this can help in assessing their eyesight, balance and liveliness.

The veterinarian will often palpate, or feel over the rabbit joints, to see if the animal has difficulty moving them. She will also look at the rabbit's claws, and if these are too long, she can cut them. If the owners are unsure, the veterinarian or nurse can teach them how to do this.

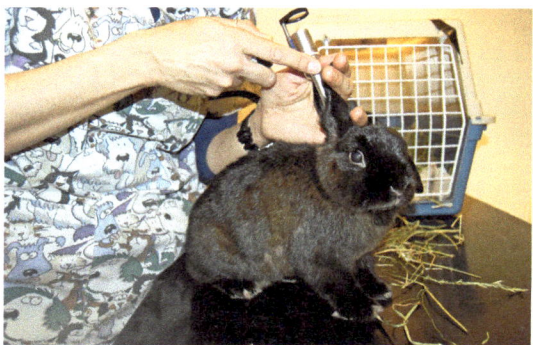

Petter's ears being examined. Photo courtesy of Marit Emilie Buseth, Norway

Ears

The rabbit's ears should be examined routinely. By feeling the ears, the veterinarian may get an indication of whether the ear canal is thickened, if it is tender, swollen, plagued by wax or has an infection. For mild to moderate cases the veterinarian may give eardrops or other treatment. If middle ear infection is suspected, X-rays of the rabbit's head are recommended. This is normally done when the rabbit is under sedation or anaesthesia. More advanced imaging such as CT (computerized tomography) is also beginning to be used.

An otoscope/auroscope may be used to examine the ear, and this is normally well tolerated by the rabbit, but may not be long and narrow enough to fully visualize the base of the ear. For that, an endoscope may be preferred, but this is usually only possible under sedation or anaesthesia. Samples of discharge, pus, crusting or other abnormalities may be obtained for further examination.

Rabbits with lop ears are more prone to ear infections. If your rabbit is scratching its ear or shaking its head frequently, it should be investigated.

Nose

Rabbits breathe through their nose and the vet will want to examine whether the nose is clean, fresh and dry. This may involve visualizing the nostrils with a bright light, or even passing an endoscope into the nasal cavity under anaesthesia. Rabbits are very clean animals, and may have removed any discharge from the nose with their front feet, so these should also be examined.

Eyes

The veterinarian will examine whether the eyes look normal, if they are of equal size and position and if there is discharge or congealed matter around the eye. Liquid and pus in the corner of the eye is often an indication of the rabbit having blocked tear ducts, a condition that is common in rabbits with inflammation of the incisor roots. Using a lacrimal duct canula, it is possible to flush these clean with a physiological saline solution and eventually test the discharge for bacteria. This may be done under general or local anaesthesia. The veterinarian will also perform extra tests if they suspect damage to the cornea, often involving the use of specific dyes placed in the eyes to highlight scratches and ulcers.

A full ophthalmic examination may take a considerable amount of time, specialized equipment and a darkened room.

Harald had an ulcer on his eye. Photo courtesy of Marit Emilie Buseth, Norway

Teeth

Dental problems are perhaps the most common reason why rabbits become ill. Signs are not always visible before the teeth have developed a chronic condition, and the incisors may be completely normal in a rabbit with even severe, painful changes in cheek teeth. Rabbit-experienced veterinarians may feel advanced changes in the jaw, as well as being able to see and assess the foremost premolars using an otoscope/auroscope. The vet cannot see all the teeth in the mouth of a rabbit while it is awake, so if the rabbit is showing symptoms of discomfort, one would therefore recommend a thorough dental check under general anaesthesia.

A holistic dental check involves general anaesthesia, visual examination of the oral cavity by means of a speculum or head-positioning device intended for rabbits, radiographs in different positions and an examination of the tear ducts.

Some rabbits must receive corrective dentistry, periodically for the rest of their life, while most acquired dental problems may be reversed, slowed or at least managed by treatment and a proper diet (see Chapter 6).

Skin

Rabbits may have fur mites and other parasites without showing any symptoms. The vet will investigate the coat and skin to evaluate the state of skin health. Skin scrapes, fur plucks or samples obtained using sticky tape applied to the skin and fur may be helpful. The vet will also go over the rabbit's body, looking for abnormalities such as abscesses, tumours, infected wounds or thickened skin, and look for signs of solidified faeces or moisture in the fur. On detection of any abnormality, necessary treatment can be implemented. In tricky cases, biopsies may be required, which generally require at least local anaesthesia. In some cases, a useful sample can be obtained by placing a small needle into a mass, to obtain a few cells for microscopic examination. This is less traumatic but often does not reveal as much information as a true biopsy.

Heart and lungs

The veterinarian will also listen to the rabbit's heart and lungs. In relation to body weight, rabbits have relatively small lungs, making them particularly vulnerable to pneumonia and other respiratory ailments. Rabbits do not tolerate draughts and have to live in sheltered and pleasant temperate places.

The vet will listen to the rabbit's heart and hear whether it beats steadily, regularly and strongly. Subtle changes may not be detected because of the extremely rapid rate of the rabbit heart compared to humans and other, larger pet animals, but if it is suspected that the heart is enlarged or another heart or lung disease has developed, taking X-rays of the chest cavity will be recommended. Other tests such as ECG (electrocardiography) and ultrasound (echocardiography) may also be carried out, although this may require more specialized equipment, and your rabbit may need to see not just a rabbit-experienced vet but one with particular experience and expertise in the field of cardiology.

Abdomen

The veterinarian will palpate the abdomen to gain information on the stomach and intestines. They will ask the owner about the rabbit's droppings, their size

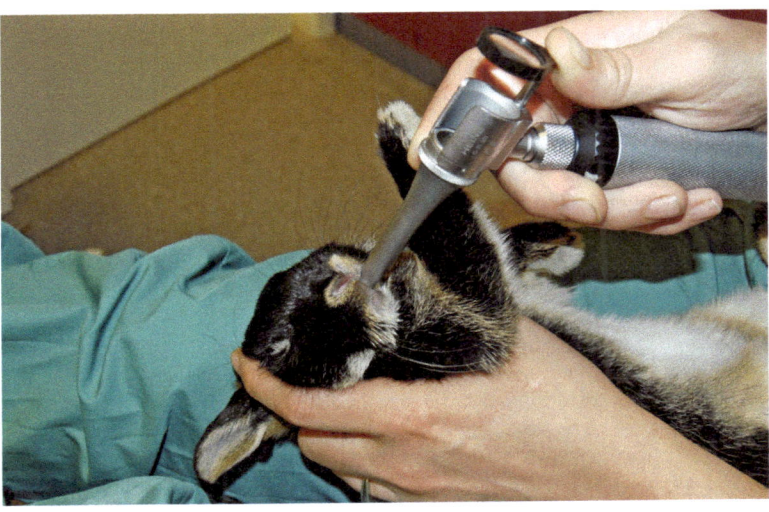

Dental examination of Melis. Photo courtesy of Marit Emilie Buseth, Norway

and shape, and if any stool is available, it could be looked at under a microscope to see whether the intestinal flora is normal or not. It is best to examine the caecotrophs, but a trained eye can receive information from the usual dry pellets as well. Parasites and other harmful organisms could also be detected, and a start could be made with medications. It is helpful to bring such a sample or at least be able to note and describe any changes in the pellets over the past few days, or even to bring photographs.

The veterinarian will also take the rabbit's temperature. Reduced temperature can mean shock or hypothermia, which can go together with decreased activity of the intestine, and treatment must be initiated immediately. Elevated temperature usually means a generalized infection, which could arise from any organ system and need not be located in the gastrointestinal tract, although heatstroke and overwhelming stress can also elevate the body temperature. Rabbits have a normal rectal temperature of from 38.2 to 39.8°C.

If a rabbit with a serious injury or illness is brought to the veterinarian in a hurry, the consultation will develop slightly differently. In such emergency cases, first aid (fluids, pain relief, oxygen therapy) will be given to stabilize a debilitated rabbit before a further diagnosis can be made and specific treatment initiated.

What Do I Do to Keep My Rabbit Healthy?

Life expectancy can be said to be anywhere from 8 to 13 years. However, average lifespan is unfortunately much lower since most rabbits suffer from malnutrition, little or no exercise, poor health and an inability to perform normal rabbit behaviour. While the potential lifespan is 13 years, a thorough survey revealed that the average lifespan of pet rabbits is approximately 4.2 years, due to poor husbandry.[3]

So what can be done to increase the chances of a long and happy life?

The best one can do to prevent health problems in rabbits is to give them a high-fibre and grass-based diet, provide sufficient freedom of movement, as well as offering socialization with a suitable companion.

Receiving the required vaccines and having regular check-ups with a rabbit-experienced veterinarian increases the chances of your rabbit staying healthy and reaching its optimal age. Myxomatosis and viral haemorrhagic disease (VHD) are two serious infectious diseases of rabbits that are preventable by vaccination.

Neutering is the only preventative or prophylactic procedure carried out on rabbits. Females run a great risk of developing uterine cancer if they are not neutered. Males are at risk of testicular cancer if not castrated, but this is much less common (read more about neutering in Chapter 7).

The company of others also seems to have an emotionally protective effect during stressful situations, and rabbits living in pairs or groups are shown to be less susceptible to stress than those living alone. When we know that stress might lead to a number of disorders, it is understandable that solitary housing in general appears to reduce the rabbit's lifespan as well.[4] Rabbits are very social animals and should be offered suitable company, preferably that of another rabbit. Having a rabbit living outside alone should be avoided, so if one has no chance to provide proper housing for two rabbits, one should rather let the unaccompanied animal move inside, so that the rabbit can be a companion rabbit within the house and consequently enjoy attention and social enrichment. Social deprivation is unfortunately a major welfare challenge for many pet rabbits.

Rabbits require both hiding places and lines of sight with escape routes to feel comfortable in an environment, and it is important to offer an adapted area to avoid stress and anxiety.

Poor air quality is less of a concern with rabbits cohabiting with humans than those outside in small confined areas. Ammonia odour is released as a result of infrequent cleaning of a hutch or litter tray, and together with dust levels from shavings this may lead to respiratory disorders. Smoking appears to be a risk factor as well and should be avoided in the house.

Senior Rabbits

Rabbits will, like other species, develop age-related ailments. Their sight and hearing can be impaired, joints and muscles will be stiffer and the rabbit will need to rest more than before. It is debatable when a rabbit can be considered senior, as some show signs of ageing as early as 4 years of age, while others may be youthful until they are 11. What is important is to allow for the rabbit's needs. A blind rabbit will most likely resent you rearranging furniture in the room, as it relies on being secure in the environment. A rabbit with some stiffness in the body may not jump up in a litter tray with high edges, and overall the senior companion animal will need a little extra warmth and peace.

Brain ageing is little understood in rabbits, compared to humans and even other pets, but is likely to affect learning, memory and behaviour. While other physical and behavioural reasons should be excluded, in rabbits displaying altered mentation like loss of litter training or being disoriented, this may be a cause of such problems.

> Early treatment of various diseases can be crucial for survival. It is therefore important that attentive and adult people are familiar with the rabbit's subtle language and immediately respond if signs of distress are noticed.

Symptoms of Pain and Illness

Rabbits do not complain. They do not whimper when breaking a rib, they do not cry when having a stomach-ache, and they will not call for help when struggling with intense pain in the mouth. As prey, they will try to hide weaknesses and vulnerabilities, and anyone living with rabbits must know how to see through this, ideally by knowing the normal behaviour of the individual rabbit extremely well.

House rabbits have an advantage compared to their counterparts out in garages and gardens, since one more easily keeps track of a rabbit's behaviour and habits when living together. Thus one is also more aware of the rabbit's subtle signals that something is wrong, which means that one can start with any recommended treatment as soon as early warning signs are observed.

Symptoms or injuries requiring an emergency visit to the vet

The symptoms listed below require immediate action and urgent treatment by a veterinarian.

- Anorexia; not eating.
- Lack of stools.
- Bloated stomach.
- Depression; losing interest in the surroundings.
- Trouble breathing normally.
- Grinding of teeth.
- Inability to urinate.
- Collapse.
- Seizures.
- Fractures.
- Heat stress.
- Myiasis – flystrike (see pp. 101–102).
- Diarrhoea in infant rabbits.

Other symptoms of pain and disease

The symptoms mentioned below require the rabbit to be seen by a veterinarian at the earliest convenience; in addition to examining the rabbit, be aware and provide any necessary first aid in the meantime.

- Not running around like normal; not wanting to be in favourite spots; not begging for treats; or could be hiding.
- Other changes in behaviour, such as becoming aggressive or nervous.
- Isolating itself; withdrawing from both you and companion rabbits.
- Immobility.
- Changes in the nature, quantity or size of droppings.
- Eating less or cutting out specific foods.
- Not being able to lie down comfortably.
- Lack of grooming, fur loss or itchiness.
- Excessive grooming and licking on a wound or body part, indicating pain in the area.
- Wet snout.
- Sneezing.
- Blood in the urine.
- Weight loss.
- Increased water intake.
- Increased urination.
- Paralysis.
- Moisture of the rear end as a result of urine or faeces stuck in the fur.
- Diarrhoea, unless in young rabbits (see above).

> **Help, my rabbit is not eating and I see no droppings. What shall I do?**
>
> Rabbits are dependent on a constant food intake and become easily dehydrated when going off their food and water.
>
> Rabbits that have not eaten or passed faeces during the last 24 hours must be taken to the veterinarian as an emergency. However, you can provide first aid earlier by:
>
> - giving oral fluid;
> - giving nutritional support (see below);
> - contacting your veterinarian and following the instructions.

From Snout to Tail

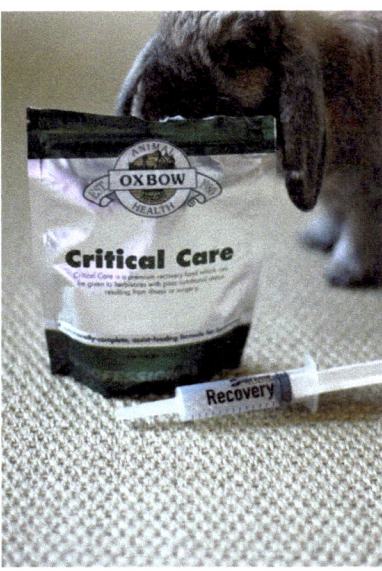

First aid kit. Photo courtesy of Marit Emilie Buseth, Norway

A rabbit pharmacy should include:

- Special diets for syringe feeding, e.g. Critical Care from Oxbow Animal Health, Recovery from Supreme, or Emeraid.
- Disposable syringes. Can be purchased at pharmacies. Useful sizes for feeding include 1 ml, 2 ml, 5 ml and 10 ml. Cutting the tips off can make it easier to administer high-fibre foods.
- Probiotics, e.g. Fibreplex from Protexin.
- Digital thermometer with a soft plastic tip, ideally one that registers temperature rapidly, e.g. in 30 seconds, to avoid excessive periods of restraint.
- Heating blanket, heating pads or a hot water bottle. Avoid electrical heat mats, which may be chewed.

Rabbit Pharmacy and First Aid

If one has a rabbit one must also have a Rabbit Pharmacy. Rabbits can be acutely ill and require immediate nursing. First aid provided on site may be crucial and decisive for the animal's recovery, and correct knowledge and a well-equipped pharmacy will often be vital for the rabbit's survival. You must, however, be aware of the local laws regarding the use of medicines, and their administration, in your country. You should establish a good relationship with your veterinary practice and, due to such legal restrictions in some countries, they may not be able to supply medicines for use without seeing your rabbit(s).

Nutritional Support

Nutritional support and a supply of liquid are the most important aspects when nursing a rabbit.

Assisted feeding may be essential for the animal's survival and recovery, and it is therefore important to be aware that situations where this is necessary can suddenly occur.

While cats and dogs can survive for days without food, rabbit digestion will be reduced when it has been more than 5 hours without nourishment and can cause serious illnesses if it goes on too long.

Critical Care from Oxbow Animal Health, Recovery from Supreme and other nutritional support mixes are nutritionally complete powder that when added to water becomes a porridge one can give to herbivores not eating for themselves. It is an ideal mix of fibre, protein, vitamins and minerals, and it is designed to stabilize and speed up rabbit recovery. The Fine Grind version of Oxbow Critical Care has a smaller particle size, and is easier to give through smaller tipped syringes, but larger fibre particles may be better for encouraging gut movement.

If you do not have these products in the house, one can make a mashed pelleted diet of rabbit pellets mixed with water. Diets must be soft enough to pass through a syringe. Liquidizing or blending them can help to make them easier to syringe. Other options include canned pumpkin, alfalfa powder and liquidized vegetables.

Baby foods are not recommended, as these often contain sugary fruit, potatoes with starch or dairy products, which the rabbit cannot tolerate and that can exacerbate an already imbalanced bacterial flora in the digestive tract.

How to give nutritional support

Make a mixture of powder and warm water in the proportions directed on the package. Place the resulting food material into an appropriately sized syringe (see below for amounts). This may be achieved by opening the syringe and filling it from

> Assisted feeding should take place as indicated by a veterinarian, as there are some cases where syringe feeding should not occur until a diagnosis is made, for example when the rabbit has a complete blockage of the digestive tract, severe facial trauma, and respiratory disease where the rabbit may inhale food.

Be careful when providing nutritional support. Offer the rabbit only 1–2 ml porridge each time, remove the syringe from the patient's mouth after each injection so the rabbit gets the opportunity to chew and swallow. A suitable total food intake is 30–50 ml/kg of the rabbit's bodyweight, over a 24 hour period. This should be divided into a number of smaller feeds, from hourly, to every 4–6 hours, depending on the level of assisted feeding the rabbit requires.

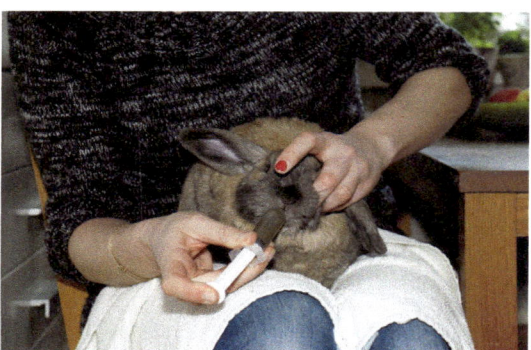

The author giving Even critical care. Photo courtesy of Aksel Hunstad, Norway

> **Checklist for everyone who has responsibility and care for a rabbit**
>
> **1.** Do you know which veterinarian to contact in an emergency and how to get there? Who should you contact if the rabbit needs emergency assistance outside opening hours? This may be a different centre to your normal veterinary practice.
> **2.** Do you have a rabbit pharmacy as mentioned above?
> **3.** Do you have a rabbit carrier available?
> **4.** Do you have clean towels you can pack your rabbit into when needed?
> **5.** Are your rabbits insured?

the wide-open end, or sucking it up through the smaller tip end. It is preferable to use a syringe with a large tip, but if you only have ones with small openings, it is possible to cut off the tip with claw clippers. This makes it much easier to draw up thicker food material.

Keep the rabbit on your lap with the rabbit facing away from you, towards your knees. Insert the syringe into the side of the rabbit's mouth, just behind the large front teeth. Rabbits are not predators and consequently have no canine teeth. The gap between the incisors and molars is therefore a perfect place to insert the syringe.

Alternatively, one can place the patient on a safe surface that is not slippery, either on a table or on the floor with you. Get someone to help you keep your rabbit calm, or wrap it in a towel so that only the rabbit's head is protruding. Such a 'rabbit burrito' will help you to keep your companion animal steady and reduces their stress levels. You can either do it yourself or get assistance.

A rabbit should never be syringe fed when it is lying on its back. It may suffocate or get liquid food in the lungs, which would be fatal.

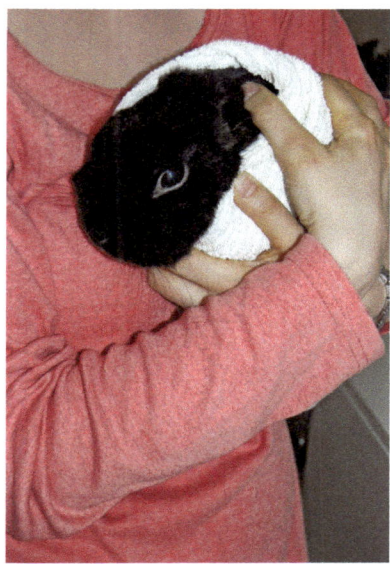

Rabbit burrito. Photo courtesy of Marit Emilie Buseth, Norway

> ### Alternative methods of giving medicine
>
> Some find it difficult to give medication orally to rabbits. A good tip is to add drops of liquid pain relief drugs, which are often very palatable, antibiotics or other medications on a basil leaf and then give it to the rabbit, which hopefully will eat it. Basil has a leaf shape that allows the droplets to remain. Make sure there are not too many drops on each leaf, and test if the method works on your rabbit in advance so you do not spill the medication. Dandelions are another option, as the hollow stems may hold liquid medication.

Fluid therapy

Fluid therapy is often the most important factor for stabilizing an ill rabbit.[5] Since the species masks clinical signs of diseases and injury as long as possible, the rabbit will often be significantly dehydrated when the owner finally detects that something is wrong.

All owners should therefore be aware of the rabbit's subtle signs of discomfort and pain, in addition to having equipment to provide oral fluid as immediate first aid. Use the same methods and types of syringe as for assisted feeding, as explained above.

For the rabbit to receive necessary fluid subcutaneously or intravenously a veterinarian must be consulted.

Remember that the all fluids must be given at the rabbit's ideal body temperature.

Temperature

It is important to take the rabbit's temperature when it is ill. Your veterinarian can show you how to do this.

The temperature should not normally be below 38.2°C. If it is below 37.6°C there is evidence of hypothermia, which may indicate the onset of shock and impaired gut motility, and it is very important to keep your rabbit warm. Wrap it in blankets or make sure to raise the temperature in other ways.

Avoid further stress and keep the surroundings dark and quiet.

> Petter had very sensitive intestines and consequently suffered from a number of problems associated with his digestion. I suspected that something was wrong when he didn't storm towards me in the morning, when he sat moping behind the couch, when he isolated himself and didn't want to be with his companions or if he refrained from jumping up on to his favourite place on the piano to have his siesta. In addition, if he didn't run around uncontrollably like a lightning bolt when offered papaya drops, I had to examine him to see whether something was wrong. These were early warning signs that something was amiss.
>
> One morning Petter did not come running towards me as usual; I found him sitting hunched up under the couch. He was cold and gave a clear expression of being in severe pain. I tried to entice him forward with fresh herbs and papaya treats but he sat motionless. He would not be with his companions, and I saw that he had trouble lying down. He certainly was not comfortable and I had to react immediately.
>
> He had just been eating and behaving normally. When I had gone to bed the night before, he ran around and left behind nice droppings as well. Now he was sitting there, in his own urine, being depressed and lethargic.
>
> I immediately took him to the clinic, and he was stabilized with fluid under the skin and pain relief. A heating plate was used to keep him warm, and contrast dye X-rays were taken of the gastrointestinal tract and teeth. The radiographs revealed gas in the digestive system, but there were no visible signs of blockage in the intestine. We proceeded by giving Critical Care quite frequently and went home in the evening.
>
> Despite a recommended high-fibre diet it was unfortunately not the first time he had been ill, so as I had done before I gave him Critical Care every 3 hours as well as additional fluid and pain relief when necessary, and made sure that he had a pleasant temperature. The digestion was slowed and it took a long time to speed up. Both the rabbit and I were tired, I took time off work and gave necessary treatment around the clock, and little can surpass the relief I felt when I noticed the first dropping – after 3 days I finally saw a dropping. It was a tiny one but still a dropping.
>
> When he eventually chewed on a tunnel of unpilled willow, I finally relaxed. Petter had made it!
>
> I continued to assist his feeding some days, and he gradually ate more and moved about.
>
> Having a diseased rabbit is like having a sick child. It requires care and surveillance even at night.

Petter was ill and lost weight, but survived thanks to intensive care. Photo courtesy of Marit Emilie Buseth, Norway

Diseases

Rabbits suffer mostly from diseases caused by an undesirable lifestyle. Poor diet, lack of mobility, social deprivation and unsanitary conditions cause most rabbit health issues. Fortunately it is easy to prevent most health problems by ensuring proper living conditions, but some rabbits are genetically prone or otherwise unfortunate enough to develop diseases that require surgery or other medical treatment.

Diarrhoea

Problems and diseases associated with the digestive system are so important in rabbit medicine that they will be discussed in a separate chapter (see Chapter 6).

Diarrhoea is also an example of a condition that in most cases is the result of improper feeding, while in certain cases it may also be caused by infection. Parasitic gut problems are uncommon in rabbits, with the exception of coccidia *Eimeria* species usually affecting young animals and causing liver and intestinal coccidiosis. Coccidiosis infection is most prevalent in areas where rabbits live together and unhygienically as the infection is transmitted through droppings that are more than 2 days old. Symptoms of infection include diarrhoea, an enlarged or bloated abdomen, weight loss, poor coat and depression. Treatment is fortunately usually successful, if commenced before permanent intestinal or liver damage has been caused, and before death due to dehydration from diarrhoea.

If diarrhoea persists for a few days on just hay and water, the vet must be consulted.

Gastrointestinal stasis

Almost the opposite problem is a slowing down or complete halt of normal gastrointestinal movement. This manifests itself in a lack of or reduction in food intake and faeces output. This may be sudden in onset, or gradual. It may be complete or partial, and it may be mild or extremely severe. In some cases this is due to a physical blockage of the gut. In most it is due to a slowing or cessation of the rabbit's gut movements through the inhibiting effect of pain or stress, or a poor diet. Blockages with tumours, abscesses or foreign bodies can prove rapidly fatal and require emergency veterinary attention to diagnose and treat.

Read more about gastrointestinal stasis and other digestive ailments in Chapter 6.

Respiratory system

A number of conditions can lead to respiratory problems, and many of them are related to the rabbit's life situation. Poor cleaning of litter trays leads to a concentration of ammonia, which will irritate the respiratory tract; the same applies to cigarette smoke, various household sprays and other air pollution. Obesity, heatstroke, foreign objects in the nose, physical and mental stress can also be contributing factors, as can an impaired immune system, which reduces resistance to outbreaks caused by bacteria, parasites or viruses.

Respiratory disease is not commonly found to affect wild rabbits. It is likely that significant respiratory disease impairs a rabbit's ability to escape predation, and they therefore do not suffer from anything more than transient disease.

Snuffles is a respiratory infection in rabbits and describes a condition involving sore noses, animals sneezing, inflammation of the nasal mucosa and runny nose and eyes. Often one can observe wet paws and faces of the rabbits trying to groom themselves and thus rubbing discharge on to the face and paws.

'Rabbit snuffles' is often caused by an overgrowth of the bacterium *Pasteurella multocida*, the most common cause of respiratory infection in rabbits. Other bacteria also associated with rhinitis in rabbits are *Bordetella bronchiseptica*, *Staphylococcus aureus*, *Moraxella catarrhalis* and *Mycoplasma pulmonis*. These bacteria are often found in the airways of apparently healthy rabbits, but can lead to severe infection of the upper respiratory tract in rabbits with impaired immune systems as a result of physical or mental stress. Outbreaks are very contagious and seem to be most prevalent in large farming systems or other places where many rabbits live together in cramped conditions, particularly where ventilation is poor.

Signs of snuffles include:

- sneezing;
- runny eyes; and
- nasal discharge.

An infection of the upper respiratory tract will spread further from the nose to the eyes, ears and eventually the lungs. Treatment should be initiated as early as possible because the disease is difficult to treat in the later stages. Bacterial culture testing of respiratory secretions, ideally via a swab placed deep inside the nose, must be taken to discern the types of antibiotics that kill the particular bacteria. Anaesthesia is required to take a deep nasal swab for culture.

Treatment is usually centred around appropriate antibiotic therapy, based on cultures, as above. Treatment should continue until approximately 10 days after the end of all symptoms, as otherwise recurrence is likely. Treatment of in-contact rabbits may also be necessary, as they may have milder forms of disease and infectious organisms present. If there is little or no response to treatment within a week, the diagnosis, or treatment, should be re-evaluated. Other useful treatments include inhaled antibiotics and antiseptics, drugs to open up the airways and help clear mucus, and anti-inflammatories.

Urinary tract problems

Urinary tract problems are extremely common in rabbits and can be difficult to treat. This is because a number of factors can cause these problems, and all of them must be identified and treated.[6]

The most commonly presenting problems are incontinence and urine scalding, and 'sludgy bladder disease'. These are problems at each end of the urinary disorder spectrum, with either too little or too much urine being produced at times.

Kidney or bladder stones (urolithiasis) and sludgy urine are two common urinary tract disorders, which are often the result of a diet relatively rich in calcium, insufficient water intake and a lack of exercise. The rabbit will produce thick and white urine, of almost toothpaste-like consistency at times.

Cystitis is an inflammation of the urinary bladder that may cause frequent urination, a burning sensation, and may be associated with sludgy urine. The disease is often caused by a bacterial infection,

Harald was suddenly in a bad mood and soon he sneezed multiple times with white purulent pus coming out of his nostril. We took him to the vet who gave him broad-spectrum antibiotics. However, this did not seem to work, as Harald continued to sneeze in addition to his nose being clogged up. It was exhausting for him to breathe and he became snappy towards his companions as well. When he had difficulties eating, we ensured he received sufficient nutrition by assisted feeding of Critical Care.

We went to the veterinarian again and during general anaesthesia a swab sample from his nose was taken. His tear ducts were flushed and cleaned, and that must have been a relief as they were blocked with pus. The nose was also gently cleaned, and it is important to be aware that rabbits must always be fully anesthetized and intubated when this is done, otherwise liquid may end up in the lungs. An X-ray of his teeth was taken and Harald was awakened.

Both blood tests and X-rays proved good results, and when we finally got the answer on the culture test they showed that it was the bacterium *Pseudomonas* that ravaged his nose.

We therefore changed antibiotics and acquired a nebuliser.

After a month of three daily visits into the nebuliser and frequent medication, he thankfully seemed to get rid of his symptoms.

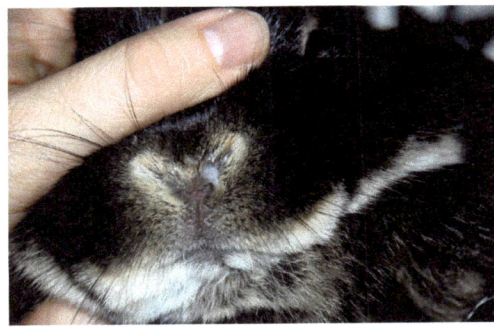

Harald with a runny nose. Photo courtesy of Aksel Hunstad, Norway

although few bladder diseases in rabbits are primarily or solely caused by bacteria, so antibiotics are, at most, only a small part of the treatment plan, which needs to look at the whole picture.

Neurological origins, such as trauma caused by injury, or neurological infections such as *Encephalitozoon ciniculi* are common causes of urinary incontinence. An unbalanced rabbit might be reluctant to move around, and head tilts and hind-leg problems may cause urinary incontinence. Rabbits with some form of hind-leg stiffness, immobility or pain might be unable to use a litter tray. Retention of urine may also be a factor in the development of sludgy urine problems. (Read more about neurological infections on pp. 105–106.)

Urine scalding is a problem often seen in rabbits that have lived in a damp environment, either due to humid and dirty surfaces in a cage, or homeless rabbits going outside in a wet climate without the ability to keep dry. Passing urine whilst not in the correct position also allows urine to dribble down the inside of the legs and underside of the tail. The rabbit's urine is corrosive and an irritant, making the skin extremely sore if in contact, without the animal being able to clean itself, and may cause inflammation and fur loss in rabbits that are unable to remain dry.

A rabbit reluctant to clean itself properly may also develop accumulation of caecotrophs. The constant wet fur and inflamed skin, in addition to the retained caecotrophs, makes the rabbit vulnerable to flystrike as well.

So how to treat urinary tract problems in rabbits?

Urinary tract problems may be primarily bladder/kidney problems, or secondarily due to musculoskeletal problems, as above. Therefore, a diagnostic approach that includes imaging the spine and pelvis and assessing mobility is required initially. It is also important to ensure that problems are not arising from the genital tract, which is most common in unneutered females with uterine tumours. Specific diagnostics will then include blood tests for kidney disease, ultrasound examination of the kidneys and bladder, and X-rays of the bladder. A fresh urine sample is helpful, ideally one obtained by 'free catch' rather than from the litter tray, which will be contaminated.

Simple bladder infections can be treated with appropriate antibiotics, but there are generally other factors, such as a high volume of calcium-rich irritant sediment material excreted into the urine and incomplete bladder emptying.

Anti-inflammatories may help with both bladder discomfort and any underlying stiffness issues. Other drug therapy used may include medication to improve urine output or decrease the calcium content of the urine.

Physical flushing of excessive sediment, either by catheterization or surgery, may be needed.

A diet with lowered calcium levels may help to prevent recurrence, but only if this is actually thought to be the problem.

Oxbow Animal Health has developed a high-fibre supplement containing various beneficial ingredients to support the overall urinary health of rabbits. 'Oxbow Natural Science Urinary Support' is recommended in rabbits suffering from urinary tract problems, alongside other specific treatments as necessary.

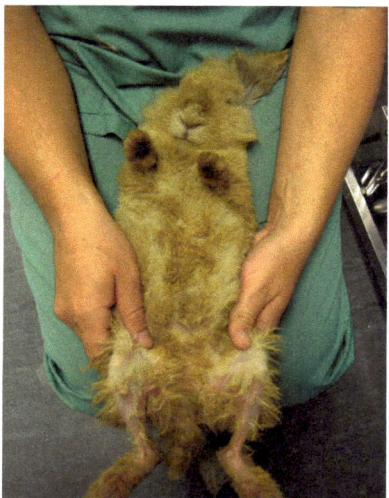

Trixie had urine scalding. Photo courtesy of Smådyrakutten Lillestrøm, Norway

> Trixie was rescued in the summer. She had been fending for herself at a housing estate for months before a passer-by finally contacted Dyrebeskyttelsen, the Norwegian animal charity. The neighbours said she was doing fine but she was worn out. She instantly fell asleep in the carrier, and upon closer inspection we saw that she was missing most of the coat on her hind legs, the stomach and under the tail. The skin was red and inflamed, and she naturally enough had severe pain and had been struggling to keep herself warm. Trixie received treatment and could fortunately rest, eat well and get a companion in dry and safe living conditions. The fur grew out again, and Trixie got herself a rabbit companion.

Abscesses

Bacterial skin infections will often result in skin infections and abscesses. Breaks in the skin can become infected, or underlying ailments can lead to both internal and external abscesses. Rabbit pus is thick and forms solid tumour-like lesions that are difficult to get rid of by simply emptying them, as is often done in dogs and cats. The entire abscess must be surgically removed with subsequent treatment. Bacterial culture tests should be performed from the abscess capsule, as the pus itself is usually bacteria free. An antibiotic resistance test (bacterial culture and sensitivity testing) is often helpful to find

Skin diseases

Ectoparasites are parasites that live on the surface of their host, e.g. mites, lice and fleas. Different species of ectoparasites live on the rabbit's skin, with only one species, the ear mite, *Psoroptes cuniculi*, living inside the ear canal. In uncommon cases this parasite can spread out on to the skin of the head, neck and genitals. Other types of ectoparasites can live in various places on the rabbit and produce variable pain and symptoms.

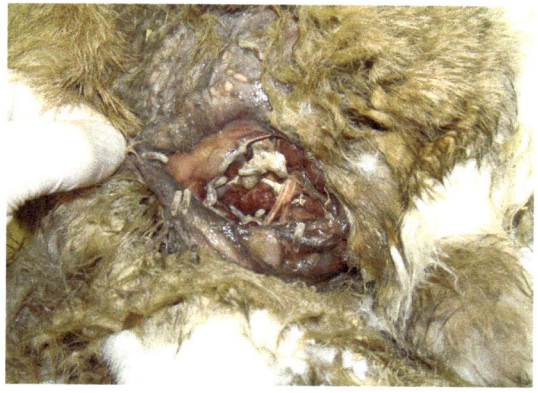

Flystrike. Photo courtesy of Glen Cousquer, UK

out what antibiotic is likely to have the best effect on the treatment. However, that often requires more invasive sampling than simply obtaining pus, as pus is generally composed of dead bacteria.

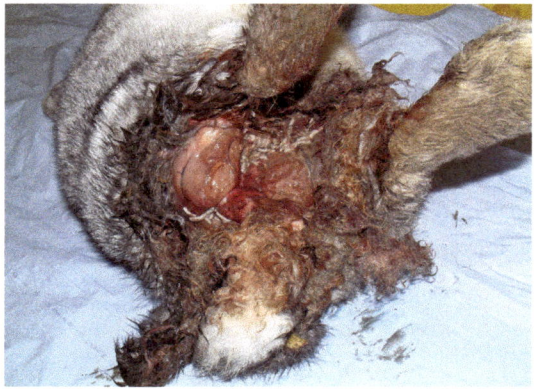

Flystrike. Photo courtesy of Glen Cousquer, UK

Flystrike or fly larvae attacks

Flystrike is a dreadful condition that causes enormous suffering to the rabbits who are attacked. Myiasis, as the disease is also known, often affects rabbits kept outside on damp surfaces during summer, but can affect even indoor rabbits during warm and especially warm and humid weather. It is the common fly *Lucilia sericata* that is considered to cause the disorder.

Especially prevalent is the problem of rabbits that are afflicted by diarrhoea, urine scalding, retained caecotrophs or other moist faeces, and consequently developing a dirty abdomen, whereupon the flies lay eggs on the skin of the rabbit. The soiling of hindquarters and damp surfaces will attract flies and possible attacks, and in the heat the eggs transform to larvae in 2 hours, and they eat themselves into the rabbit, away from the light. This naturally causes extreme pain for the rabbit, who is unable to remove the maggots.

Rabbits need to be examined around the rear end three times a day during warm periods. Moisture in the litter tray and coat must be avoided, and to ensure a dry and clean litter tray this must be changed every day. If your rabbit lives in a cage with litter or sawdust, this must also be kept dry. Upon detection of the onset of an attack, it is of major importance to respond immediately and seek a rabbit-skilled veterinarian, since the rabbit needs immediate emergency treatment to survive.

Regular observation allows early identification of the problem. If there are only fly eggs present, these can be removed by the owner, although it is important to also clean the rabbit of any faeces and urine, and to arrange an urgent examination by your veterinarian to look into the reasons behind the attack before they cause other problems, or flystrike occurs again.

If there are maggots present, or if the rabbit has open wounds or is ill in any other way, then immediate veterinary attention is required. Pain relief, fluid therapy and nutritional support are required, and a prompt diagnosis as to the underlying cause(s) and treatment with antibiotics are needed.

Maggots and fly eggs must be removed from the skin and from any wounds. This is straightforward on unbroken skin and superficial wounds. However, when maggots are found under the skin, in pockets at the edges of wounds, or in the anus or prepuce of vulva, removal is more complicated.

Individual removal using forceps is often the most certain way of avoiding missing any maggots, although eggs may be brushed away from the animal using a small brush, such as a toothbrush.

Maggots are attracted to warmth, and many find the gentle application of warm air from a hairdryer to be helpful in tempting them out. Gentle flushing using sterile warm fluids may help in removing them; however, this should be from deep to superficial in order to avoid driving them deeper.

Various chemical options are suggested for prevention and treatment of flystrike. Permethrin containing products, in spot-on or spray form, with added germicidal components may be used. Insect growth hormone regulators, that prevent the hatching out or metamorphosis of eggs to larvae, are also used as a preventative. Care must be taken to avoid leaving maggots behind, especially in body cavities, as they may continue to burrow deeper, or if they die, they may provoke a foreign-body reaction within the rabbit's body. For this reason, it is always advised that rabbits are thoroughly re-checked at least once for maggots, approximately 30–60 minutes after the initial removal process.

Once the condition has been treated, it is extremely important to determine the causal factors in that particular rabbit, to treat the underlying problem(s) and to prevent future attacks. Typical causes are obesity, dental disease, musculo-skeletal problems, abdominal pain, urine scalding, diarrhoea or retained caecotrophs, and a diet too high in simple carbohydrates and calories and too low in fibre.

Preventative treatment of all the above is vital, and correction of any of the above factors helps to avoid future flystrike.

Flystrike is challenging to treat, and it is easier to prevent attacks than to cure it. Any rabbit may be affected, although it mostly strikes rabbits living outdoors. To prevent disease, it is important to stick to a good diet so that droppings are dry and firm.

Sore hocks – pododermatitis

Sore hocks or bacterial pododermatitis is another skin condition that mainly arises from the rabbit's living conditions, but also as a result of fractures or other injuries in the rabbit, which can cause undesirable pressure. Obesity, long nails, confinement, living on wire floors or other rough substrates, and urinary incontinence are some of the reasons that often lead to suffering. Hairless areas under the feet with subsequent red skin are the first signs of a rabbit developing sore hocks. Early treatment and improvement of living conditions is important. Untreated pododermatitis will lead to pain, infection in hairless areas, and eventually spread to the underlying bone and cause blood poisoning.

In the wild, the rabbit's claws will sink into the grass or other soft ground. This helps the foot to keep contact with the surface and consequently avoids unfortunate pressure. This is the case both when they are moving and sitting still, and to prevent suffering due to sore hocks it is important to be aware of the rabbit's anatomy and needs.

In captivity, rabbits are often provided soiled and dirty bedding, rough or hard surfaces, poor nutrition and reduced access to exercise and natural movement, which all make them vulnerable for developing pododermatitis. When the claws are unable to sink into the ground, the front paw is being raised. This leads to an unnatural burden on the rabbit's back legs, and the increased pressure on the hocks may result in damaging sores.

Prevention is the best treatment.

Skeletal and muscular disorders

Osteoporosis

Rabbits have a very light and fragile bone structure, which makes them vulnerable to caging conditions with no or little potential for normal activities, such as running, hopping or even just sitting up on their hind legs. Rabbits living a sedentary life develop a poor bone structure, and those who have to spend their lives in a typical cage often suffer from thinning of their bones. In rabbits developing such osteoporosis, the brittle bone is being exposed for fracture and damage. Since the rabbit is excellent at hiding pain, it is hard to know whether your

Ekorn, the rabbit with the silky and shiny coat.
Photo courtesy of Aksel Hunstad, Norway

Ekorn had been kept under unacceptable conditions. The rabbit had the most beautiful silky coat I have ever seen, but when he was brought to my house I immediately noticed that he was unable to put down his rear feet, and when I turned him around I discovered horrible sore hocks.

I installed him in my office, where I covered the floor with thick and soft fleece blankets. I had hay covered with towels in his litter tray to minimize the pressure on his sore feet. Then I put on a soothing cream and ensured that he was eating.

Further treatment was to provide antibiotics, pain relief and fluids.

We grew close to each other in the ensuing weeks, as I spent a lot of time caring for him, and finally his feet recovered.

He was a wonderful and cheerful rabbit, and a perfect example of a personality one can never forget. I am sure that everyone involved in rescuing rabbits has some individuals with a special place in the heart, a little rabbit that they can never forget.

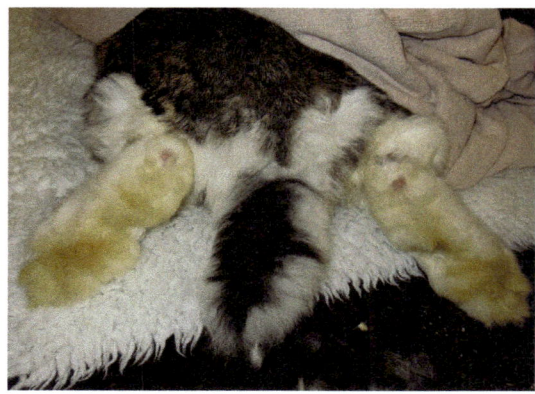

Sore hocks. Photo courtesy of Richard Saunders, UK

companion rabbit actually has a breakage or other abnormalities in the skeleton.

The best one can do is to prevent such problems by providing a sufficient and facilitated living environment.

Rabbits that have been housed in standard cages will be prone to injuries and fractures due to weakness in the bone structure. Because of cage dimensions with no opportunity for exercise, these rabbits will lack coordination, muscles and flexibility that are normal for the species.

In addition to exercise, rabbits are also dependent on a sufficient amount of calcium in their diet. Young animals in particular need a suitable, well-balanced diet to build up a healthy bone mass, and a low calcium diet is a contributing cause of a fragile skeleton.

Lack of vitamin D, an unbalanced compound feed and lack of sunlight also predispose to this disorder.

Osteoarthritis is correspondingly widespread among rabbits, and is a type of calcification arthritis. A degenerative change in the articular cartilage, bridging between the bones, and further proliferation of surrounding bone tissue leads to reduced mobility and subsequent pain in the affected areas.

Compared with dogs and cats, very few rabbits are treated for chronic pain due to osteoarthritis. It does not mean that rabbits are less susceptible, but shows that they are good at hiding pain, in addition to the lack of tradition of treating sick rabbits.

Lameness in dogs is often revealed by an abnormal gait, and this is something most people recognize. A subtle change in the way the rabbit jumps or moves on the other hand is not as easy to detect. Few people have an extensive knowledge on how rabbits naturally run and bounce around, but the vast majority know what a healthy dog looks like when it runs. It is particularly difficult to assess sedentary rabbits in cages.

A much overlooked symptom is that rabbits with calcification in the back, knees and thighs may have trouble grooming themselves on the back and find it difficult to keep themselves clean around the anus and genitals. The exposed rabbits will often resist treatment as being held often amplifies the pain.

A sedentary life in a cage often leads to obesity and osteoarthritis in joints. Weight control, sufficient movement and well-tempered climate will prevent and to some extent improve the situation of those already poor rabbits.

Cold and damp climate both trigger and worsen the chronic pain associated with osteoarthritis. Exposed rabbits may be better off staying indoors in the chilly periods. In 2007, researchers at Tufts University in Boston reported that every 10°C drop in temperature corresponded with an incremental increase in arthritis pain. Increasing barometric pressure was also a pain trigger in the Tufts study.

Oxbow Animal Health has developed a high-fibre supplement containing various beneficial ingredients to support the overall joint health of rabbits 'Oxbow Natural Science Joint Support' is recommended in rabbits suffering from skeletal and muscular disorders, alongside specific treatments as necessary.

Fracture and pain are common in rabbits, usually after a fall or injury when twisting and wriggling out of a grip. A child may be in danger of losing a wanton rabbit on the floor; a rabbit kicking can easily injure himself if the owner is attempting to hold it firmly; and rabbits can fall down from the table, be squeezed in doors, etc. Rabbits want to camouflage any damage, and even with fractures they will still pretend that everything is in perfect order.

Our rescue centre has received several so-called 'angry' rabbits. Upon further health examinations we often find that they have a fracture in the ribs, a broken tooth or other typical fall and crush injuries.

Rabbits will try to hide weakness and pain, but they will also have to defend themselves against any pain and danger. If your rabbit has unpleasant memories of being carried by anyone, it will try to prevent this from happening again. A natural defence from a frightened rabbit in pain will often be misinterpreted as aggression.

Many rabbits suffer from having lived under cramped conditions. It is important that these rabbits

Ten years in a hutch

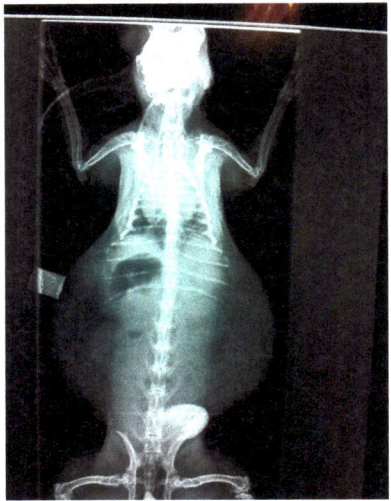

A rabbit after 10 years in a hutch. Photo courtesy of Marit Emilie Buseth, Norway

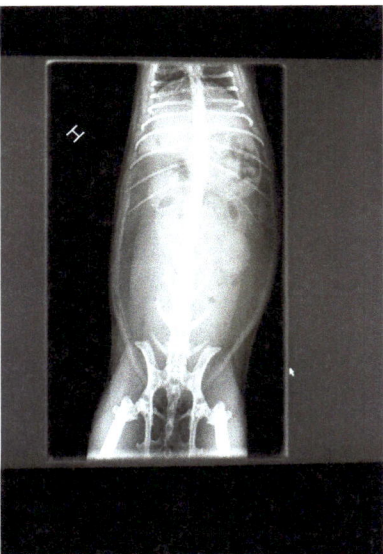

A free-range rabbit, 9 years old. Photo courtesy of Marit Emilie Buseth, Norway

When Erling finally was rescued, he could hardly walk. For 10 years he had been kept in a hutch in the garden. Initially it had probably been fun for the children, a pet they could take care of. However, the joy did not last, and Erling was soon left alone. He sat, slept, ate and sat. However, when the kids were grown up, the parents got tired of feeding him once a day, and they gave him to some neighbours who knew of a better home.

Radiographs revealed degenerative arthritis in his spine as well as his knees, hips and shoulders, incisor overgrowth and a sludgy bladder. This had probably caused pain and discomfort for several years. However, the most obvious was nevertheless his lack of abdominal muscles. Since the rabbit never had developed and maintained a normal musculature, there was nothing that kept the intestines and stomach in place. Radiographs of Erling looked more like a run-over frog than a rabbit.

He was already neutered and was introduced to his new rabbit companions right away. They were free-range in a house, but while the two others were running around, Erling did not know how to. He tried to walk, but it was obviously both painful and unfamiliar, he lost his balance and spent the first few days resting.

Despite the challenges, Erling seemed grateful for his new life. His caretaker left herbs around the room to encourage movement, and after impressive progress he could move more easily after a couple of weeks. He was humping in a distinctive way, but he soon learnt to lift his legs off the ground while out exploring. He became more inquisitive and active each day that passed, and after some weeks he could lie completely stretched out during rest. He loved to be close to his companions, and was never tired of being groomed by his conspecifics.

Erling touched us all. Photo courtesy of Marit Emilie Buseth, Norway

Continued

> Continued.
>
> As an example of how a sedentary life injures the species, I wrote about Erling and showed revealing pictures on my blog (http://www.maritemilie.com),[7] and the old rabbit soon got to be a Scandinavian celebrity. Rabbits and poor husbandry were suddenly in the spotlight, and at some point an article about him was the most shared news online. Erling also took part in a TV show, where I in addition talked about the mental suffering, loneliness and endless boredom due to a hutched up existence. Rabbits were in the news, and people seemed shocked by how we treat animals we claim to love.
>
> Erling got to be a symbol, and I am sure his story has helped other rabbits. Sadly enough he was not going to experience a happy life for long, and one day he fell asleep and quietly passed away. Erling got to be 10 years old, but he only *lived* for 7 weeks.
>
> Please be aware; tell if you see a rabbit in distress and offer to help.

> Many rabbits have fractures in their toes. Two of mine have broken toes and were both revealed by a loss of appetite and altered activity, and confirmed by radiographic images. They received suitable pain relief and a bandaged foot after the injuries.
>
> It is important to ensure that the rabbits have claws of the proper length, to avoid damage caused when a claw becomes stuck. There is always a risk of infection with a broken claw.

get used to a bigger living environment so that they have an opportunity to train themselves to avoid damage. Rabbits that normally live in a hutch should not participate in rabbit jumping competitions, which many youngsters are interested in. Such activities could easily lead to overload and damage to a normally sedentary body. A rabbit needs daily exercise and movement, not the obstacle course weekly or once in a while.

'Splayleg' is an inherited disease in which rabbit kits are not able to position the legs. It mainly affects the hind legs that are being thrown out awkwardly behind and to the sides of the body. There is no treatment for the disease.

Illness and injuries to muscles, joints and bones can also cause paralysis and neurological disorders.

Neurological disorders

The body usually maintains a sense of balance and position as a result of the vestibular system. The vestibular system is part of the central nervous system, and an injury in this area could lead to a change in the perception of balance in the affected rabbit. The most well-known symptoms of neurological disorders in rabbits are head tilt, circling, hind limb weakness and loss of balance. Possible neurological symptoms can also include collapse, urinary incontinence and convulsions.

The most prevalent causes appear to be an infection, either the parasite *Encephalitozoon cuniculi*, or bacteria such as *Pasteurella multocida*, which causes otitis (ear infections), but it can also be caused by irritation and secondary infections due to the ear mite *Psoroptes cuniculi*. Other triggers include spinal damage, tumours, cancer, viral infections, abscesses, vascular diseases and various poisonings, and it can be difficult to make a correct diagnosis. One will therefore often choose to medicate for both *E. cuniculi* and bacterial otitis in cases of head tilt if the cause of the symptoms is not obvious. Diagnostic tests such as blood tests, X-rays and even CT or MRI imagining can be useful.

Rabbits infected with *E. cuniculi* can also develop renal failure, which can result in reduced appetite, increasing thirst and weight loss. The parasite can also affect the eyes, leading to cataracts and rupture of the lens. This may lead to inflammation within the eye and blindness.

The above-mentioned neurological incidents can also cause seizures and it is then important to try to minimize the possibility of further damage. Seizures usually last for 1 to 2 minutes and may be experienced as dramatic. Make sure to secure the area so that the rabbit is not subject to injury such as fractures whilst seizuring, talk quietly and wrap the rabbit into a soft, clean towel and go to the vet immediately if the seizures last for more than a few minutes. The veterinarian can then stabilize the rabbit.

Signals that the rabbit has a neurological disorder can come on gradually or occur suddenly. It is, in any case, naturally important to seek veterinary assistance as soon as you notice that something is wrong with your rabbit.

Encephalitozoon cuniculi is a microscopic parasite that lives inside rabbit cells, typically in the kidneys and brain. Over 50% of rabbits in the UK are positive on blood tests for exposure to the parasite, and they are most likely infected before birth or in early life through spores in infected urine. The majority of the infected rabbits will remain healthy, while a few of them will show symptoms of the disease. Antibodies to *E. cuniculi* can be detected in a blood test. However, as antibody levels may remain high for months or even years after exposure to the disease, it is difficult to know if the clinical symptoms are actually caused by *E. cuniculi* or whether another disease is involved in those animals with a positive test result. A negative result, however, means that the rabbit's current symptoms are not due to *E. cuniculi*.

A few rabbits with neurological damage may improve without treatment, as the brain finds new ways to compensate for the damaged areas. Treatment with the drug fenbendazole, however, has shown promising results and is now generally the only medicine that is used in cases of suspected *E. cuniculi*, even though anti-inflammatory drugs may also be indicated to reduce inflammation affecting nervous tissue. Treatment for 28 days has become the standard treatment in the UK.

Many rabbits respond quickly and successfully to the treatment. However, the disease is challenging to cure, and a successful treatment will not always lead to resolution of symptoms, as brain damage may already be irreversible. It is worth mentioning that a rabbit with a moderate head tilt may be perfectly functional. There are many examples of rabbits that have adapted to the condition and are otherwise healthy, and it appears that the most important prerequisite is that they continue to be active as normal.

Supporting your infected rabbit

During the disease, your infected rabbit will probably feel unbalanced or even dizzy and be reluctant to move around. The rabbit might feel that the world spins around, and will therefore need food, water and a litter tray within easy reach.

A rabbit with head tilt might have problems drinking due to the angle of the head. During the illness it is sometimes necessary to support the rabbit with fluid and nutritional therapy, while modification of the environment may be necessary in cases where head tilt persists, for example, altering the position of water bottles if they are unable to reach them.

Many rabbits with impaired balance will lean to one side. Make sure to provide the rabbit with a leaning object and try to make them comfortable during the outbreak.

Encephalitozoon cuniculi is a potential zoonosis, i.e. it can infect people with compromised immune systems.

Even has fought against *E. cuniculi*. He was unable to walk; he rolled over and lacked control over his body. After some days with Panacur he recovered, and after a couple of weeks he was like normal. Photo courtesy of Marit Emilie Buseth, Norway

> ### A rabbit with Encephalitozoon cuniculi
>
> Reidar's caretaker found him lying in his own urine in the morning. His rabbit companions sat nervously in a corner and wondered what was happening, and it looked quite dramatic. Reidar had rolled over and was not able to keep his balance and sit upright. He had also developed a severe head tilt during the night. Luckily he was taken to a rabbit-friendly veterinarian immediately.
>
> Blood samples were taken to see if he was a carrier of the *E. cuniculi* parasite; he was given fenbendazole and was sent home to continue the treatment for 28 days.
>
> His caretaker was worried. Her rabbit rolled over, she had to make sure he was stable and also ensure that he received nutrition and hydration through assisted feeding.
>
> She could not believe that he would be any better, but the next day she already noticed improvement. The second day he almost managed to sit upright, and we were quite optimistic. The progress continued, and Reidar was improving day by day. His caretaker and her husband took turns being home with him the first weeks, when he had to be fed regularly, and after 12 days he seemed normal and content.
>
> He did however have a relapse, and was both lopsided and unsteady in the body for some days. A further treatment with Panacur led to improvement within a couple of days, and he remained on the medicine for 28 days once again.
>
> He has been healthy since, and today it is almost 3 years since he rolled over and scared his family. Reidar is now 10 years old and still enjoys his life.

Cancer

Cancer, or neoplasia, is perhaps not as common, with the exception of uterine adenocarcinoma in female rabbits (see Chapter 7), as in many other species of domestic mammal. However, it is entirely possible that this is because rabbits often do not live to more advanced ages, and because diagnostic biopsies are less commonly performed in the rabbit than in cats and dogs.

Skin neoplasia

Lesions on or just under the skin are very common in rabbits. Only a small proportion are cancerous, with most being due to parasites, fungal or bacterial infections. The main 'differential diagnosis' is an abscess, a finding that could be mistaken for cancer. However, a range of benign and malignant skin tumour types have been reported in rabbits.

Internal masses

Internal tumours are not uncommon in rabbits, but the vast majority are uterine, which are amongst the most preventable of cancers. It is rare that an internal tumour becomes so large that the first sign of it is a lump that can be felt. Most often, there will be other clinical signs.

Bone tumours are uncommon in rabbits. In the vast majority of cases, any swelling associated with bone on the face is due to tooth disease and subsequent bone infection. It can sometimes be challenging to differentiate tumours from infection: radiography and laboratory analysis of bone samples may be necessary to be sure.

Uterine tumours are of course the most common tumour of un-neutered female rabbits. Depending on breed and age, they can be found in anything up to about 80% of female rabbits.[8,9] They can cause abdominal pain, blood in the urine and gut stasis. They are preventable by neutering relatively early in life. Surgery to remove the uterus can be successful in treating the problem, unless the tumour has spread to the lungs or other sites such as bone, or if it has spread within the abdomen to other organs. As a general rule of thumb, if there is no spread within the abdomen, then further spread is unlikely. However, lung X-rays are strongly advised.

Testicular tumours are not uncommon in rabbits. They typically occur in older males. In most cases, the tumour does not spread from the testicle, and castration removes the tumour in its entirety.

Mammary tumours are not uncommon in the rabbit and are most common in does over 3 years of age. Many are benign, but some are highly malignant.

Kidney tumours occur uncommonly, and may be benign or malignant. They may occur early in life or later. They are usually only found following extensive imaging of the abdomen, or incidentally at surgery. They may be almost completely insignificant, or very malignant, with few cases being in between.

Lymphoma is the second most common tumour of rabbits, and the most common tumour of young rabbits. Multiple organs are often involved, including the kidney. Affected rabbits may have a range of vague non-specific signs. Duration of illness can be from 1 week to 10 months. Enlarged lymph nodes are not commonly encountered in rabbits for other reasons, and so is a very significant and grave finding.

Liver tumours are relatively uncommon compared to the frequency in cats and dogs.

Rectal papillomas may appear as raspberry-like lesions emerging, sometimes intermittently, from the rectum. Whilst they can frequently bleed, they do not appear to be particularly painful, and do not seem to predispose to flystrike, although this is obviously a concern. They may be removed surgically.

Treatment options

Surgery is often completely successful alone. It is important to take enough skin and tissue round the edges of masses, although in most cases this will not need to be particularly extreme. In cases where wide surgery is necessary, special techniques such as skin grafts may be used.

Chemotherapy and radiotherapy has not been evaluated thoroughly in rabbits, although regimes may be extrapolated from humans and dogs/cats.

Palliative treatments, including antibiotics, steroids and non-steroid anti-inflamatories, as well as opioid painkillers where necessary at the end, may be helpful. Euthanasia is unfortunately the only option in advanced cases or those that are affecting the rabbits' quality of life.

Viral haemorrhagic disease

Viral haemorrhagic disease (VHD), also known as rabbit calicivirus, or rabbit (viral) haemorrhagic disease (RVHD), is a rapid onset, and frequently fatal, disease of rabbits.

The disease is rapidly fatal in unvaccinated susceptible animals. There may be no prior warning signs before death, and the body may be completely free of external signs, although often there is blood from the orifices (mouth, nose, anus and genitals). On post-mortem, there is nearly always substantial internal bleeding, usually originating from the liver.

Rabbits under 10–12 weeks do not seem to get VHD. No-one knows why! Animals obtain antibodies either through the placenta or through the milk, but that does not seem to be the reason that young rabbits are protected from VHD, it appears to be some other mechanism.

The incidence of this disease is often underestimated by veterinary surgeons and owners alike, as, unlike other diseases with a more prolonged course, such as myxomatosis, the animals are rarely seen to develop the disease and may be found dead with no signs, and never taken to a vet.

One study has shown a link between vaccination against VHD and lifespan, and this may reflect a higher than realized incidence of VHD, or that owners vaccinating against VHD take other important steps to care for their rabbits, or both.[10]

The virus is very resistant to deterioration in the environment, and has spread throughout the world via inanimate objects. In colder temperatures it is especially persistent, and may last for months off the host animal.

Wild rabbits are also less likely to be seen with the disease, as again they die rapidly and predators will take their bodies. The signs of VHD in wild rabbits are more likely to be mistaken for that of trauma from cars or predators in any case.

This, coupled with the ease of accidentally carrying the disease from infected fields or rabbit premises, makes it potentially deadlier than myxomatosis. There are many reports of mass die-offs due to outbreaks in rescues, breeding, showing, or sales animals, or large groups of pets.

Vaccination is therefore a vital step in prevention, with excellent biosecurity also advised in at-risk situations such as known local outbreaks. Waiting until the disease is known to be nearby before vaccinating is *not* advised, as it may be too late to afford protection at that point. Biosecurity measures include avoiding bringing in wild-grown plants from areas to which wild rabbits have access, and use of foot dips or changes of footwear after walking in potentially infected fields.

Treatment is not usually an option, as the disease strikes so quickly.

> At the time of writing, vaccination is neither necessary nor a legal requirement in some countries, i.e. Norway, due to the lack of particular diseases. In other countries it is strongly advised but not legally required (e.g. UK and USA). Contact your vet for advice if you are unsure what applies in your country.

> **Mosquito and flea control**
>
> Vaccination is the mainstay of myxomatosis and VHD prevention. However, no vaccine is 100% effective, and myxomatosis may also be avoided by flea control and screening of other household pets, where they are at risk of picking up rabbit fleas and ticks. Flying-insect prevention methods may include chemical deterrents and insecticides, and fly screens on doors, windows and outdoor enclosures.
>
> Steps to reduce the mosquito population locally may include emptying standing water or covering it with solid covers, or a thin layer of vegetable oil to suffocate larval forms. Biological control methods (e.g. *Bacillus thuringiensis*) are also suggested to reduce the vector populations.
>
> VHD is highly persistent in the environment, up to 150 days at room temperature, and may be brought into the indoor environment by humans or animals on the feet. It may also be brought in on wild-picked food material.

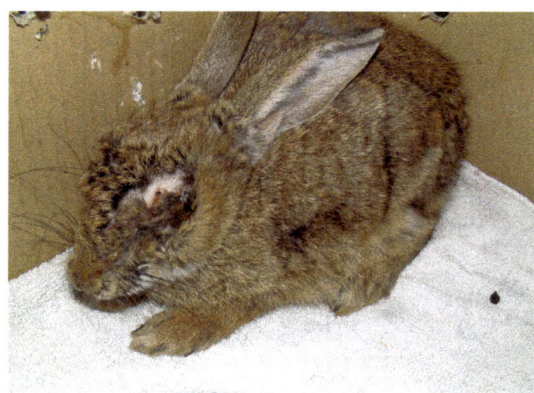

Myxomatosis. Photo courtesy of Glen Cousquer, UK

> A new strain of RVHD has been noted in continental Europe and has been designated RVHD2.[11] It is unclear how effective the current vaccines are against this new variant, although they appear to offer at least some protection, and the disease strain appears less commonly fatal than RVHD1. At the time of writing, confirmed cases of this virus have been identified in the UK, with the first such cases in spring 2014, and cases have been identified in Norway in July 2014.

Myxomatosis

Myxomatosis is a severe and dreadful viral disease, which poses a threat to both wild and domesticated rabbits. The disease will cause immense suffering and death for unvaccinated animals, and all pet rabbits should be vaccinated in countries and areas where the virus is present. The disease is usually spread by biting insects, and rabbits living outside are more vulnerable than house rabbits, even though all pet rabbits are at risk and must be vaccinated.

Myxomatosis is a poxvirus, originally a pathogen of the South American wild rabbit species with which it has co-evolved and is not generally fatal. In the naïve European species it was extremely pathogenic, and when introduced to the UK in the 1950s it decimated the wild rabbit population.

It is a distinctive disease, with the typical, or 'full-blown' form characterized by thickening and swelling of the skin of the ears, eyelids, lips and genital and perianal tissues. This gradually develops from day 5 until, in the very latest stages, the eyes are completely shut and often there is secondary bacterial infection developing under the eyelids, with pus oozing out from them.

For wild rabbits, death is typically by predation at this stage, as the rabbit is effectively blind and unable to smell. Alternatively, they may wander into roads and be killed by cars, or be picked up by passers by and taken to a vet to be euthanased.

In the pet population, the course of the disease is unfortunately similar, although shorter, as the signs are so distinctive that most owners seek veterinary attention early on, usually resulting in euthanasia.

Attempts to treat this full-blown form are generally without success. Where the rabbit has previously developed some immunity, e.g. via the parents (there is evidence of both maternal and paternal immunity) or via vaccination, the disease may be less severe, if it is contracted at all, and may sometimes take the 'atypical', 'skin myxo' or 'cutaneous myxomatosis' form. This causes coin-sized raised scabs, generally over the face and genitals, and occasionally over the skin of the shoulders and neck. These generally fall off, leaving cratered areas of raw tissue underneath, which usually fill in and heal without problems, but may become infected. They may cause problems if they involve the skin of the eyelid, nose or lips, and cosmetic surgery may be needed to stop them causing damage to the eye or preventing feeding.

Full-blown myxomatosis is nearly 100% fatal. In the wild, a small number of rabbits survive infection and pass on immunity, or they would become extinct. However, this is not of great reassurance to the individual pet owner, and although heroic measures may increase the chances of survival, and it has certainly been shown that significantly raising the ambient temperature reduces mortality, this small chance of survival has to be weighed up against the huge potential for suffering with this disease.

The cutaneous form is survivable, with around 10–20% mortality at most. In the majority of cases, cutaneous myxomatosis is self-limiting, and easily managed with simple treatment as necessary.

Bayliscascaris procyonis

The roundworms of racoons and skunks can migrate through the nervous system of rabbits if their eggs are ingested. Severe damage to the brain may result, with signs similar to that of *E. cuniculi*. Their eggs may remain infective for up to a year and are spread to rabbits when their food is contaminated by infected racoon or skunk faeces. The larvae then migrate into the brain and cause damage there, resulting in tremors, balance problems and weakness, head tilt, circling, which can lead to falling over and rolling, paralysis, seizures and collapse. Treatment is only partially effective and needs to be continued life-long. Prevention is ideal, by stopping access to potentially contaminated food and bedding.

Obesity

Obesity is a huge problem in rabbits. Although one study and review suggests that fewer rabbits are affected than individuals of other species, the consequences for rabbits are even more of a welfare issue than in dogs and cats.[12] Foot health problems, urine scalding, caecotroph retention and flystrike may result, as well as problems with mobility and cardiac health that may be seen in other species.

General Anaesthesia and Nursing

Before and during surgery

Previously rabbits did not receive necessary operations or treatments as a result of the high risk associated with rabbits being under anaesthetic. The species were neither neutered nor treated for serious dental problems as a consequence, but an increased knowledge of rabbits as well as developments in anaesthesia have fortunately changed this former trend, and rabbits of today will enjoy vital procedures in line with other animals.[13]

Preparation and examination of the hospitalized rabbit can be said to be important to the outcome of the anaesthetic, and after-care and nursing of the convalescent determines how the animal does afterwards. One must be aware of what one has to think about both prior to and after surgery.

Three main problems associated with rabbit anaesthesia are stress, hypoxia and pre-existing diseases. Additional concerns include heat loss during and after surgery, lack of nutritional support, and technical problems with maintaining access to the circulatory and respiratory systems.[14]

It is essential that the rabbit feels calm and secure prior to the anaesthetic. It should be waiting in a quiet place and not in a waiting room with predators, people, loud sounds and other things to watch out for. Rabbits are a prey animal and will be unnecessarily frightened and alert due to stress, which will lead to high adrenaline levels, which in turn may prevent sedation, as well as slowing gut movements. One should also try to avoid stress due to pain, transport, rough handling and others making the rabbit vigilant.

It is important to note that rabbits do not need to be fasted before anaesthesia. They are unable to vomit and should be offered their regular hay until half an hour prior to sedation. Being a herbivore means that they are dependent on the constant presence of food in the stomach. Be careful, however, that there is no food in the mouth or throat before providing medication.

Hypoxia, an insufficient or reduced supply of oxygen, is another threat when rabbits are under anaesthesia. Rabbits have poor lung capacity and are therefore subjected to oxygen deprivation. Being nasal breathers makes them vulnerable to blocked airways and one should monitor the patient's oxygen supply during the procedure. Breath holding is a common problem when rabbits are anaesthetized. It is a particular problem when rabbits are anaesthetized using gas anaesthesia only. This may be provided by putting a mask over the rabbit's face or placing them in a box full of gas. This method is not ideal, without appropriate pre-medicant drugs, as it causes significant stress. Rabbits may attempt to kick out or otherwise escape, injuring themselves, and even if restraint is excellent and safe, they may hold their breath. Not only does this deprive them of oxygen, but it increases the concentration of carbon dioxide in the blood, alters the acidity levels of the blood, and may eventually end in unconsciousness, or a sudden deep breath, taking in too much anaesthetic.

Pre-existing diseases such as respiratory infections, gastrointestinal issues and dentition causing pain predispose to cardiac arrest. If possible, one should stabilize sick and dehydrated animals prior to anaesthesia, by providing analgesia, fluids and a quiet room.

When premedication is given, the rabbit should be able to rest in a dark and quiet place, e.g. in a carrier covered with a blanket. One should not stand around the rabbit waiting for it to fall asleep. Being a watchful prey animal, rabbits naturally try to prevent sleep when having spectators, and this alert tendency might counteract the effects of anaesthesia. In such cases one can easily provide too much anaesthetic.

Premedication is usually a mixture of sedatives and pain relievers that should be given prior to the anaesthetic. A synergistic effect occurs, making it possible to offer less anaesthetic when premedication has been provided. This is understandable, knowing that a physiologically stressed rabbit will excrete adrenaline and have a high level of stress hormones in the blood. This might counteract the anaesthetic agents, making the anaesthesia more unpredictable, and one can see lack of sleep in an alert rabbit. Premedication thus seems to ensure that the anaesthetic agent has the opportunity to work optimally. When including pain relief, premedication will in addition lead to a better recovery of the rabbit when it wakes up.

Rabbits need analgesia to be effective before surgery is performed. The pain-relieving drugs will already be active when the rabbit wakes up and the pain begins. Rabbits are sensitive to pain and any experienced discomfort may lead to decreased recovery. A rabbit with post-operative pain will take an unnecessarily long time to convalesce, and the pain might lead to digestive problems and subsequent serious complications.

Non-steroidal anti-inflammatory drugs, or NSAIDs, such as meloxicam or carprofen, are widely used in rabbits, whereas opioids (such as those with the active ingredient buprenorphine) can be used in cases of severe pain that often occur after major surgery. One can use a combination of NSAIDs and opioids to make the life of a convalescent or sick rabbit much easier. The correct dosage is essential, and this must always be calculated from the exact weight, condition and age.

Because distress in itself causes gastrointestinal stasis, analgesia should always be offered when pain is suspected. The doses of drugs for rabbits may be quite significantly different to those used in other, more familiar species, such as dogs and cats. Veterinary surgeons unfamiliar with rabbits may need to consult more specialist textbooks and dosage guides, for example, those listed in the box on p. 88. In particular, the doses of painkillers are often much higher in rabbits, and doses suitable for dogs and cats may be so low as to have no useful effect.

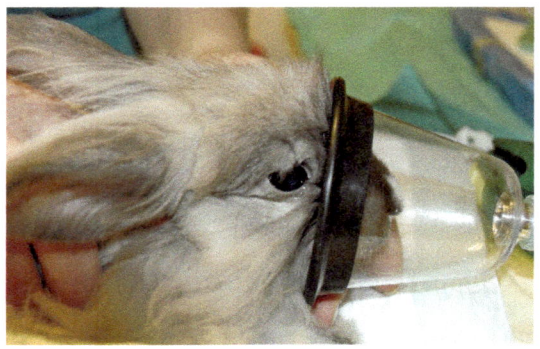

Gas anaesthesia. Photo courtesy of Marit Emilie Buseth, Norway

The veterinarian must always ensure that the rabbit's health is as good as possible before the procedure. Often, an acutely ill rabbit might be dehydrated and suffering from hypothermia (low body temperature), and it is important to stabilize the animal as much as possible before surgery is performed. The clinic should have procedures for such life-saving first aid, and trained staff should supervise the rabbit. If an apparently healthy rabbit will be under general anaesthesia, as for neutering, it is also important to be familiar with rabbit health, including taking the accurate weight. Rabbits are smaller than dogs and should, like cats, be weighed on scales accurate across the weight range 300 g to 10 kg. Veterinarians inexperienced with rabbits commonly incorrectly estimate their weights when assessing them visually, due to their different body shape to that of cats and dogs.

Being nasal breathers makes the species vulnerable to respiratory infections, and the veterinarian must allow for unknown disorders of which the owner is not aware. A rabbit might cope with an infection in the lower respiratory tract in everyday life, but when respiration is reduced under anaesthesia, secretions in the airways thicken and consequently lead to blockages. Such obstruction would lead to a lack of oxygen in the blood, known as hypoxia. It is therefore essential to offer the rabbit oxygen supply during the procedure.

One can provide anaesthesia by using gas or injectable agents. Both have advantages and disadvantages, but the right equipment and knowledge ensures the best possible results.

When using gas one has the option to adjust or reverse the amount of anaesthesia with immediate effect. Rabbits should receive sedation in advance.

Injection anaesthesia has also undergone an evolution. Former use of irreversible drugs has now given way to the reversible injection anaesthesia. By giving certain drugs one can reverse the effects of the anaesthetic. Rabbits will take a bit longer to wake up after the injection method than with gas anaesthesia, since medications need to be excreted through the body. Older rabbits with liver and kidney problems may have and less complete recoveries due to their organs not metabolizing and excreting the drugs as rapidly. Such potential problems explain why a veterinarian may advise blood tests prior to an anaesthetic, and they may modify the doses or choices of drugs in such cases.

During both procedures one can place a tube into the trachea, enabling the anaesthetist to breathe for the rabbit if this should become necessary. Endotracheal tubes are advanced into the airway under anaesthesia, and are stable, prevent against foreign material entering the trachea, and also minimize the presence of anaesthetic gas in the atmosphere, which humans may breathe in. Supraglottic airway devices are a new development, which are much easier to place than endotracheal tubes, and sit on top of the trachea instead of inside it. They have most of the advantages of endotracheal tubes and are a useful alternative method, especially in emergency situations.

Intravenous catheters can be used during both gas and injection anaesthesia. This gives the veterinarian access to provide life-saving fluids and medications as soon as possible, because the catheter is already in place in the vein.

It is possible to measure blood pressure during surgery. Blood pressure gives information about the circulation, and if it falls, one should start with immediate action in terms of giving intravenous fluids. If the pressure is dangerously low, one can provide countervailing drugs. Monitoring the blood pressure will also give information about whether a rabbit has lost much blood. There are several methods of measuring the rabbit's blood pressure, but the most widely used involves fitting a cuff around a leg or an ear under anaesthetic, much as in human blood pressure measurement.

Visual examination, listening to the rabbit and 'hands-on' assessment is of course also important when a rabbit is anaesthetized. Observations and recording of heart rate, breathing rate and pattern, the presence of reflexes, the position of the eyes within the eye sockets and the degree of muscle tone present all help to assess anaesthetic depth, i.e. whether the rabbit is dangerously near to death, or so light that it may wake up.

Rabbits under anaesthesia are prone to hypothermia. Heating the room to a suitable background temperature, placing insulating material over the rabbit and/or supplying heat by direct warmth (hot water bottles, heated bean bags, hot air systems, warming electrical pads) during and after the procedure can be crucial for the rabbit's recovery. One should also clip as little fur as possible, and use careful amounts of cleaning fluids on the skin, particularly those that evaporate, such as alcohol-based products, to minimize heat loss.

Rabbits should always be at the clinic until they are fully awake after anaesthesia. If a rabbit needs further medical care or for other reasons must remain at the clinic, trained staff should monitor the animal. It is best, once the rabbit is strong enough, for it to be provided with adequate and competent care at home, so it can stay in familiar surroundings.

One must pay close attention to a rabbit that has been through anaesthesia. Warmth and fluids are critical, and one must also ensure that digestion continues. If your rabbit will not eat after anaesthesia it is important to assist feeding. A nutritional

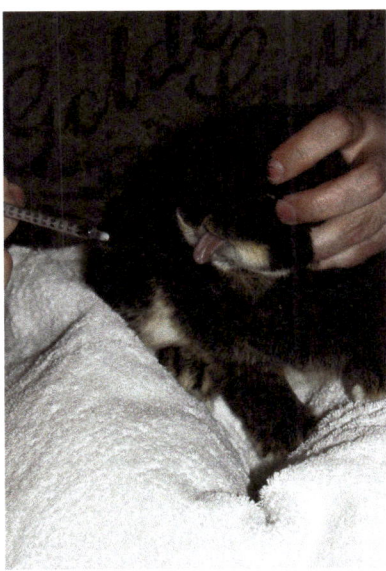

Harald tasting the medicine. Photo courtesy of Marit Emilie Buseth, Norway

support diet, like Critical Care from Oxbow Animal Health, Emeraid Critical Care system, or Supreme Recovery Diet, is a product all rabbit owners should have at home, especially after surgery. Critical Care and other products are a powder you mix into a porridge-like consistency, and it may play a critical role in how your family animals are doing in the aftermath of various procedures and periods of illness in general. The motility of the digestive system is maintained, and the rabbit will avoid dangerous dehydration and an empty stomach. (Read more about providing nutritional support on pp. 94–96.)

Owners must ensure that the patient avoids stress and keeps warm. A newly operated upon or sick rabbit cannot stay outside in low temperatures, and should be kept inside for observation and company regardless. One must of course also monitor whether there is any infection of the wound, that the rabbit eats and has normal droppings, and make sure it appears healthy. Bonded rabbits should not be separated during illness, as comfort and support are important for the recovery.

Euthanasia

If the rabbit has an incurable illness and/or injury that makes it prone to sustained pain or discomfort then euthanasia is the fairest and most humane option. This is never decided upon lightly, but equally, it is important not to keep a rabbit alive, against its own best welfare, for one's own reasons.

To euthanize an animal should be conducted as carefully as possible, and the rabbit should not feel stress, pain and fear at the end of life. Palliative opioid painkillers may be used in some cases, in the very short term, in order to make arrangements, but many rabbits will reach this point more suddenly and dramatically, and it is worth considering your thoughts on how you would wish to carry this out, if and when the time ever came. Consider who in the family, including children, should be present, whether you would like to give the companion rabbit(s) the chance to say goodbye, especially after the event, and whether you would want euthanasia to be carried out at the veterinary surgery or at your home. You should also consider whether you would like to take your rabbit home to be buried, or to receive his or her cremated ashes back afterwards.

If a rabbit has to be euthanized, the owners must emphasize that the animal shall receive a sedative injection before the lethal injection is given. Some veterinarians have, particularly in the past, only administered an overdose of anaesthesia directly in the abdomen. This is very painful and the rabbit may be stressed or even terrified. It takes several minutes before the drug has taken full effect, and the last minutes of your beloved rabbit's life will thus be filled with unnecessary pain and fear.

If you as an owner feel comfortable, you may ask to be allowed into the clinic to see that your rabbit falls asleep after the sedative injection, and that your companion rabbit, after the second syringe, falls asleep for good.

One method, which achieves a relaxed end but without being drawn out and lengthy, is, after a sedative agent is administered, to place an intravenous catheter and deliver the overdose of anaesthetic agent through this. This can be done with local anaesthetic cream placed over the vein, to further minimize pain and potential stress, and the use of a catheter ensures that the euthanasia drug may be given smoothly and effectively.[15]

Tonic Immobility

Rabbits can go into a condition where they look almost hypnotized, either when they are laid on the back or handled in some other way. It is a widespread misconception that rabbits are relaxed in such cases, when lying immovable on their back in your lap or in the crook of your arm. However, this is not the case. Tonic immobility is a condition certain prey have acquired to escape enemies. It is a defence mechanism rabbits habitually revert to when they have no other options to escape. The rabbits will play dead, so that the predator loses interest and lets the prey go.

Studies have concluded that tonic immobility is fear and stress related. Physiological measurements support the assertion that such treatment should only be carried out in appropriate situations, such as when cutting nails or during certain investigations.[16]

When rabbits do not have other opportunities to defend themselves, the hypnotized state is their only way out. A rabbit that is scared and kicks frantically when held should therefore not be held to the point of the rabbit's surrender and the wriggling stops. Rabbits under such circumstances are in a state of fear where one sees physiologically measurable stress reactions on the heart, respiration, blood pressure and level of stress hormones in the blood.

> **How to hold a rabbit**
>
> - Rabbits should never be held by their ears.
> - Rabbits should never be held by the scruff of their neck.
> - Rabbits should always be held in a firm and secure grip. Use both hands. Hold a hand under the hindquarters while supporting the body with the other. Hold the rabbit snug against your body so it does not feel like it will lose its footing.

Notes

[1] Gunn-Dore, D. (1997) Comfortable quarters for laboratory rabbits. Available at: http://www.awionline.org/pubs/cq/five.pdf (accessed 1 July 2012).

[2] Harcourt-Brown, F. (2002) *Textbook of Rabbit Medicine*. Butterworth-Heinemann, Oxford, UK.

[3] Schepers, F., Koene, P. and Beerda, B. (2009) Welfare assessment in pet rabbits. *Animal Welfare* 18(4), 477–495.

[4] Schepers, F., Koene, P. and Beerda, B. (2009) Welfare assessment in pet rabbits. *Animal Welfare* 18(4), 477–495.

[5] Saunders, R. and Whitlock, E. (2012) Nursing hospitalized patients. *BSAVA Manual of Exotic Pet and Wildlife Nursing*. British Small Animal Veterinary Association, Gloucester, UK, pp. 129–166.

[6] Saunders, R. (2010) Urinary tract problems. *Rabbiting On* 1, 4–5.

[7] Buseth, M.E. (2013) Ti år i bur og effektene av et stillesittende liv. Available at: http://maritemilie.com/2013/02/12/ti-ar-i-bur-og-effektene-av-et-stillesittende-liv (accessed 23 July 2013).

[8] Heatly, J.J. and Smith, A.N. (2004) Spontaneous neoplasms of lagomorphs. *Veterinary Clinics of North-America: Exotic Animal Practice* 7, 561–577.

[9] Hilyer, E.V. (1994) Pet rabbits. *Veterinary Clinics of North-America: Small Animal Practice* 24, 25–69.

[10] Romain, P. (2011) The effect of viral haemorrhagic disease (VHD) vaccination on the longevity of pet rabbits (*Oryctolagus cuniculus*) in Scotland. In: Roberts, V. (ed.) *British Veterinary Zoological Society Proceedings*, Cheshire, UK, 12–13 November 2011, p. 44.

[11] Le Gall-Reculé, G. et al. (2013) Emergence of a new lagovirus related to *Rabbit Haemorrhagic Disease* Virus. *Veterinary Research* 44, 81.

[12] Meredith, A. (2012) Is obesity a problem in pet rabbits? *Veterinary Record* 171, 192–193.

[13] Brodbelt, D.C. (2006) The confidential enquiry into perioperative small animal fatalities. Available at: http://www.rvc.ac.uk/Staff/Documents/dbrodbelt_thesis.pdf (accessed 10 February 2013).

[14] Harcourt-Brown, F. (2002) *Textbook of Rabbit Medicine*. Butterworth-Heinemann, Oxford, UK.

[15] Walshaw, S. (2006) Euthanasia. In: Meredit, A. and Flecnell, P. (eds) *BSAVA Manual of Rabbit Medicine and Surgery*, 2nd edn. BSAVA Publishing, Gloucester, UK.

[16] McBride, A. Trancing Rabbits: relaxed hypnosis or a state of fear? Available at: http://www.hopperhome.com/Trancing%20Rabbits-Tonic%20Immobility%20.pdf (accessed 10 May 2010).

Lovisa and Harry. Photo courtesy of Emma Almquist, Sweden

Anton. Photo courtesy of Tina Solicki, Norway

6

Rabbit Nutrition

Wild rabbit eating grass. Photo courtesy of Alana M. Scmitt, http://www.inkubus.com

Domestic animals cared for by humans should be offered a diet as close as possible to their wild counterparts. In the case of rabbits, they have a digestive system that is specifically developed to make use of a nutrient-poor and fibrous plant diet. Their entire feeding strategy and behaviour is adapted towards nutrition based on grass, and the domesticated rabbit is dependent on nourishment derived from hay, various types of grass, forbs, herbs and leaves.

Since most people do not have fields and pastures where they can collect the necessary amount of diverse grasses on a daily basis, the rabbit's needs will not be covered by just feeding hay. However, hay should always form the greatest proportion of the diet, with a limited amount of high-fibre pellets or nuggets for mineral and vitamin supplementation.[1]

Most illnesses that affect domesticated rabbits are a direct or indirect result of suboptimal environment and nutrition. Incorrect diet, a lack of access to sufficient space to exercise, inappropriate substrate, predators and other environmental problems will inevitably predispose to a variety of disorders. It is therefore important to understand the rabbit's nutritional needs and natural habitat in order to offer the rabbit a suitable life in a domestic setting. Knowledge of rabbit behaviour is also necessary to detect subtle symptoms of disease and digestive problems at early onset, which is vital in addressing these issues early enough to implement effective therapy.

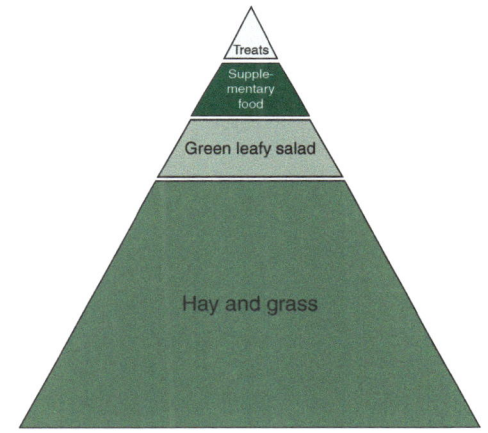

Illustration showing the ideal amounts of different foods in a rabbit's diet. Illustration courtesy of Marit Emilie Buseth, Norway

Rabbits' Diet

At least 85–90% of the rabbit's diet should consist of hay, grass and dried grass. A good guideline amount is a volume the same size as the rabbit.

In addition, the rabbits need a limited amount of high-fibre pellets/nuggets for mineral and vitamin supplementation; 20–25 g pellets/nuggets per kg ideal bodyweight is enough. This is normally only a couple of tablespoons a day.

Green leafy salad should also be part of the daily diet.

Fresh water should always be available.

Rabbit diet as illustrated in the pyramid on p. 117. Photo courtesy of Katarina Vallbo, Sweden

Hay

Choosing hay

The hay offered should be of high nutritional quality, and one can evaluate it by its feel, smell and appearance. The same holds for judging the hygienic quality, and the hay should never be mouldy or dusty.

Variety and selective feeding

Variety is the spice of rabbit life, and offering different hays will ensure both nutritional diversity and enrichment for herbivores. Rabbits, like many other herbivores, have evolved to differentiate and select high-energy items, to provide energy for growth, activity and reproduction. However, this food selective mechanism causes concerns when domesticated rabbits are offered high-energy diets or large amounts of vegetables and fruit, since the animal will consequently choose these high-energy foods above hay and grass. By providing herbivores with a diversity of hay, grass and herbs instead, one will experience a stimulated and curious rabbit that spends more time eating. Offering a variety of hay gives the rabbit the opportunity to feed selectively, which again will satisfy rabbits' innate need to search for food.

Both grass and lucerne/alfalfa hays are available, but grass hays should always be the main diet for adult rabbits, and timothy hay is the best. Lucerne/alfalfa hay contains more calcium and protein and is only suitable for growing rabbits or as a tasty treat. Due to the rabbit's unusual calcium metabolism, a diet high in protein and calcium is not advisable for adult rabbits, as this may lead to urinary calcium sludge (see p. 87 and p. 127).

Providing hay

Knowing that the rabbit is a selective feeder, as described above, explains why it is necessary to replace hay on a daily basis. Uneaten old hay must be removed and replaced with a fresh supply.

Hay can for example be offered in a hayrack or in the corner of the litter tray. Note that rabbits can urinate on the hay in the litter tray, and that whilst this might be an ideal and preferred substrate to encourage the use of the litter tray, one must always make sure to provide hay for eating in an alternative appropriate way. (See Chapter 8 for different solutions.)

It is very popular for rabbits to eat directly from the bag. Photo courtesy of Katarina Vallbo, Sweden

Supplementary food

How much pellets/nuggets?

Pellets/nuggets are only a supplementary food and should only be provided in a limited amount. A quantity of 20–25 g/kg ideal bodyweight is enough, and this is normally only a couple of tablespoons a day. This means that a rabbit weighing 2 kg only needs 40–50 g per day.

Rabbits will not eat enough hay if offered excessive amounts of other foods, and it is therefore extremely important to follow these instructions. Eating enough hay is crucial for the correct chewing action and wearing down of the teeth.

Choosing supplementary food

There is a huge and varied range of food intended for rabbits, but historically there has mostly been a poor range of balanced diets on the market, which fortunately has improved recently. Muesli, mixes for rodents and grain-based diets have been marketed as proper rabbit food, while this is highly unsuitable for the species (see box on the following page). When marketing cannot be trusted, it can be difficult to know what to buy. We will therefore introduce some guidelines for how to choose your rabbits' dried food.

A pelleted diet suitable for rabbits should always be based on grass and have a nutritional composition as below:

- Fibre: >20%
- Protein: 12–14%
- Fat: 1–4%
- Calcium: 0.6–1.0%
- Phosphorus: 0.4–0.8%
- Vitamin A: 10,000–18,000 IU/kg
- Vitamin D: 800–1,200 IU/kg
- Vitamin E: 40–70 mg/kg
- Magnesium: 0.3%
- Zinc: 0.5%
- Potassium: 0.6–0.7%

Examples of these include Burgess Excel, Oxbow Essentials, Supreme Science Selective and Beaphar Nature.

There is a distinction between compressed and extruded pellets. The old fashioned compressed pellets are produced by compressing material together. They are limited in the amount of fibre, and may therefore promote uneven dental wear and digestive diseases.

Extruded pellets/nuggets, on the other hand, are made by making a mixture and then forcing it through a small hole in a dye plate to produce the desired size and shape of a pellet/nugget. As a result of the manufacturing process, fibre length can be greater than that in the compressed pellets, and is thus better for the rabbit's digestive system.

Extruded pellets may also be referred to as nuggets by some manufacturers in some geographical locations. However, if in doubt, it is important to read the packaging information, or refer to the manufacturer's website for more details.

It is also worth mentioning specific veterinary formulations, such as the Supreme Vet Care Plus range, which are physically much larger pellets, with a higher fibre content (up to 34% crude fibre) and specific added ingredients for particular clinical conditions such as urinary tract disorders, obesity and digestive disorders.

Providing pellets/nuggets

Why offer pellets/nuggets in a bowl? Based on the rabbit's natural behaviour and feeding strategy, it is more stimulating and entertaining to search for the food. It is possible to increase foraging behaviour by providing pellets/nuggets in a treat ball or by spreading them on the floor, in a basket with hay or around the room.

People with companion rabbits often think it is sad to see an empty food bowl. However, everyone with rabbits needs to know that it is in the rabbit's best interest to gain a restricted and limited amount of supplementary food such as pellets. Even though they seem to lose their head when finally served pellets, it does not mean they are starving. With constant access to fresh hay and water, they are provided with necessary nutrition. Most illnesses that affect domesticated rabbits are a direct or indirect result of incorrect nutrition, so think of wild rabbits when you give your companion rabbit food.

There is a trend to refer to rabbits, guinea pigs and chinchillas as Folivores, or Fibrevores, in order to stress their requirement for high fibre, low energy density foodstuffs, which require long periods of time spent eating.

Vegetables

It is unfortunately a widespread belief that rabbits and carrots are a perfect match. This is, however,

Troya and Gabriel are following the diet tips and are healthy and happy rabbits. Photo courtesy of Katarina Vallbo, Sweden

based on comics and pictures from popular cartoons rather than an up-to-date knowledge of the rabbit's digestive system.

Rabbits are dependent on nourishment derived from hay, various types of grass, forbs, herbs and leaves, and the sugar content in non-leafy vegetables and fruit is much higher than in any food of their natural diet. The species' digestive system is not designed for eating sugar and starch found in carrots, apples or grains, and many digestive disorders could be avoided if one provided the rabbit a natural diet.

As nutritional variety and supplementation, one will often provide a higher level of diversity by

---

Muesli versus pellets

Supplementary dried food may take the form of pellets/nuggets or a multicomponent mix, commonly referred to as 'muesli' due to its visual similarity to human breakfast foods. In this chapter, we learn why muesli should be avoided, which is supported by recent studies carried out by the University of Edinburgh and FERA, sponsored by Burgess Pet Care. The purpose of the study was to see the effects different diets had on dental health and digestion in general, and their findings revealed that feeding muesli led to painful dental and digestive problems. Thorough research found that rabbits fed on a muesli diet, with or without hay, had to face major welfare problems compared to those on a more favourable hay and pellets/nuggets

diet. A hay-based diet, with small, carefully measured amounts of extruded pellets/nuggets, is clearly best for the rabbit's health and welfare.

The rabbits fed on a muesli diet became rapidly obese, they developed the first warning signs of dental disease, had slower gut motility, abnormally small droppings and more uneaten caecotrophs, indicating digestive abnormalities. They also drank less water, which is a factor that can lead to urinary tract problems.

The Animal Welfare Act 2006[2] states that pet owners have a legal duty to care for animals' needs. Pets at Home have taken this into consideration and decided to remove muesli-based diets from their shelves. We hope that other retailers will follow.

offering different types of hay and grass than a couple of non-leafy vegetables.[3] In addition, the rabbit's foraging behaviour will be encouraged.

Choosing vegetables

Most rabbit books have lists of suitable vegetables recommended for rabbits. However, easily digestible carbohydrates found in fruit and non-leafy vegetables are not beneficial for the species' digestive system and should only be offered in a limited amount.

If providing non-leafy vegetables, make sure to offer vegetables with a low sugar content, such as celery or green peppers.

Green leafy vegetables and herbs are excellent as a supplementary diet. They are low in sugar and represent a varied and tasty addition to the main diet.

Twigs and branches

Twigs are also an important step for rabbit-proofing the house. Photo courtesy of Marit Emilie Buseth, Norway

Rabbits love to have twigs and branches to enjoy. They can clean a twig of bark in seconds, and also gnaw a little now and then. This can also prevent destruction of other wood and furniture in the house.

Stay away from trees that have fruit with stones, such as cherry, dwarf cherry or plum, as these will be toxic for rabbits. Several decorative shrubs and flowers in the garden may also be toxic.

Common trees such as rowan, pine, aspen and birch are safe options. Twigs from apple and pear trees, and branches and leaves from currant, raspberry and blackcurrant are also tasty and safe to offer. Bilberry shrub is also extremely popular and a healthy treat.

Make sure the plants are not sprayed with insecticides, fungicides or other potentially toxic treatments.

Treats

Herbs are a popular and healthy treat that can be given in almost unlimited quantities. Tiny bites of vegetables are also popular. In addition, companies such as Oxbow and Burgess have tasty treats that are safe to offer rabbits.

Unfortunately, most of the sweets and treats sold and marketed for rabbits are unhealthy and dangerous and should be avoided. Yoghurt drops, different chew sticks and drops in all shapes and colours are usually very high in sugar and will potentially lead to digestive problems.

What to Avoid

'Muesli' and mixtures

'Muesli' and mixtures made for rodents and even rabbits should be avoided. They are high in ingredients that the rabbits do not tolerate, such as fat, sugar and starch, in the shape of nuts, corn, seeds and fruits. Being a selective feeder, the rabbit will in addition select and eat the most nutritious parts and consequently end up with an even more incomplete nutrition. To ensure the rabbit has a balanced nutrition, one should provide a 100% pelleted food, homogeneous pellets with no added colouring.

Dairy products

Rabbits are lactose intolerant and should never be offered any dairy products. Therefore one should not offer them yoghurt drops, even though they are sold as treats in many pet shops.

When the microbial population in the gastrointestinal system is disturbed because of stress or disease, then it may help to provide good bacteria found in dairy-free probiotics, such as ZooLac Propaste (ChemVet), BioLapis or Fibreplex (Protexin), AviPro (Vetark) or Critical Care (Oxbow Animal Health). These products will enhance the natural microbial environment, which is important to enable the rabbit to recover.

Bread

Bread and other pastries contain high levels of carbohydrates and should therefore be avoided.

Rabbits cannot tolerate and utilize easily digestible carbohydrates, and a diet consisting of sugar and starch will eventually lead to digestive disorders.

Others to avoid

Potatoes, lettuce, sweetcorn, seeds, oat, sunflower seeds and other seeds, lentils, beans, biscuits, cakes and other feeds are known to lead to inappropriate bacterial overgrowth or possible blockage of the intestine.

The following are toxic: onions, rhubarb, stems and leaves of tomato, seeds and stones of fruit, avocado and chocolate.

Young Rabbits

During their growing period, rabbits will have an increased need for a correct balanced diet, as the development of their bones, skull and skeleton demands an appropriate level of calcium and vitamin D. Rabbits younger than 6 months should therefore be offered pellets especially developed and designed for this purpose, which includes energy-dense alfalfa hay and provides stabilized nutrients for young and active animals. In addition, young rabbits can be fed a daily amount of alfalfa hay, which is perfect for young or lactating animals that need concentrated nutrition.

The amount of pellets provided to growing rabbits is calculated based on their estimated adult weight. It may be given in higher quantities for their first months, as long as they continue eating sufficient amounts of hay and their droppings are round and fibrous.

Young rabbits do not have a highly developed immune defence and may be especially vulnerable towards changes in the microbial population due to the introduction of new foods, especially at or shortly after weaning. One should therefore always make sure not to change the young rabbit's diet overnight, but continue to give the pellets it previously ate and rather change it over a period of 2–4 weeks.

Since introduction to vegetables and salads may also lead to changes in bacterial growth, one should only give these if one is sure that the young rabbit is used to them, for example if their mother ate the same vegetables during pregnancy and lactation. Research has also shown that rabbits' food preferences follow their mothers', and this makes it easier to feed rabbits healthily from weaning when their mothers have been fed on those items as well.

Pregnant and Lactating Does

See Chapter 12.

Obesity

Obesity is a huge welfare problem in rabbits. Excessive weight puts the rabbits at risk of digestive ailments, joint disease, urine scalding, sore hocks, caecotrophs retention, flystrike and mobility problems, amongst others.

If rabbits are too fat their body weight will regulate itself with the introduction of a fibrous and recommended diet, plus sufficient opportunities to exercise.

A fibrous diet will normalize the weight, as rabbits normally eat about 5% of their body weight a day. If this is made up of more nutritious commercial food, the rabbit will become fat and suffer a slow moving gut.

Burgess Excel Light Rabbit Food is high in beneficial fibre (38%) and is aimed to reduce and control a rabbit's weight.

Improving the Body Condition of Skinny Rabbits

The best one can do to regulate the rabbit's weight is to provide a recommended and fibrous diet. This is the case for both skinny and obese rabbits, since their digestive system cannot tolerate changes beyond their natural diet. Rabbits in recovery, after surgery or severe neglect for example, may instead be offered pellets especially made for young and active animals and more nutritious alfalfa hay. Probiotics may be added to the rabbit's diet, as these help to improve digestibility of food.

Changing a Diet – Introducing New Foods

Wild rabbits experience seasonal differences in their nourishment. The nutritional value in the available forage varies and the rabbits adjust to this over time. Likewise, pet rabbits should be adapted to a new food gradually, and it is important to be aware that rabbits do not tolerate rapid changes in the diet.

Rabbits are vulnerable to changes in the microbial population due to the introduction of new foods. Sudden changes in diet can upset the digestion, so the rabbit needs to become accustomed to new foods gradually.

This rule applies both to new brands of pellets, new vegetables and outdoor grazing. House rabbits with sudden access to a garden in the springtime and consequently unlimited amounts of different grasses, dandelions, leaves and other plants, may develop painful and serious digestive diseases. Be cautious and offer the rabbit some of the fresh grass gradually before allowing free access to the outdoor enclosure. Also ensure that the rabbits are provided with their regular hay at all times.

For the same reasons it is also important to introduce vegetables and a new brand of pellets gradually. One should stick to the same supplementary food and only change this if previous feeding was inappropriate. When switching to another brand of pellets, the old one should ideally be mixed with the new brand for a couple of weeks, gradually increasing the proportion of the new, until only the new one is provided.

Water

A domesticated rabbit will most likely consume less water than wild rabbits, even though pet rabbits drink more liquid water. This is because the wild rabbit's diet contains 80% water, while the water content in hay and dry pellets is about 10–15%. As a consequence the wild rabbits eat more and work harder to fulfil their nutritional needs, compared to the domesticated rabbits offered nutrient dense 'concentrates'.[4]

A fibrous diet will help regulate the rabbit's water consumption. A diet based on hay and long fibre will ensure that rabbits are drinking more water than when being fed on grain-based pellets or mixes. Compared to other animals, they have a high water intake, 50–150 ml/day/kg body weight, and a normal rabbit of 2 kg will therefore drink as much as a 5 kg dog. A higher water intake will consequently lead to higher volumes of less concentrated urine, which may prevent urolithiasis (formation of painful kidney stones) and bladder sediment or 'sludge'.

Rabbits with water in a bowl have higher water consumption than those with water provided in a bottle and a bowl is therefore recommended.[5]

Dinka drinking water from a bowl. Photo courtesy of Katarina Vallbo, Sweden

---

How to prevent digestive diseases

- Make sure that the rabbit's diet is based on hay with a high nutritional quality. Hay must always be available and replaced daily to encourage eating. This is important since rabbits are selective feeders and will choose the best parts of the hay and leave the rest uneaten. If not provided with new hay, suboptimal hay intake can be the result.
- Make sure to give only the recommended amount of supplementary pellets and other food items; this is important to ensure optimal hay intake (see illustration on p. 117 for optimal diet).
- Make sure the rabbit is provided with balanced and rabbit-friendly pellets (see p. 119).
- Fresh water must always be available. Rabbits with water in a bowl have a higher water consumption than those with water provided in a bottle.
- Make sure rabbits have adequate exercise opportunities.
- Prevent ingestion of indigestible material, such as carpet, plastic or cat litter.
- Make sure to groom the rabbit, twice daily when they are shedding.
- Avoid rapid changes in diet.

Petter, Harald and Melis eat healthy and tasty herbs – perfect treats. Photo courtesy of Marit Emilie Buseth, Norway

> **What to be aware of: early identification of gastrointestinal problems**
>
> - Any decrease or change in food consumption should be noticed and acted upon. If a rabbit cuts out various foods or completely stops eating, this can be a symptom of serious dental diseases. Having different jaw movement patterns while eating foods causes issues with the cheek teeth and incisors, and can cause various symptoms. Rabbits with spurs on their cheek teeth will for example stop eating hay while they can still consume pellets and vegetables. A rabbit with severe pain in the incisors may still eat hay while it stops eating vegetables. It is therefore important to be aware of the rabbit's changes in preferences.
> - Be aware of the rabbit's normal faeces. Changes in amount, size and consistency are often a sign of an incorrect diet too low in fibre. If the droppings are harder, looser, darker or have decreased in amount, it is advisable to feed the rabbit exclusively on hay for a couple of days. If the stool still is abnormal and unhealthy a veterinarian should examine the rabbit.
> - If the rabbit does not pass faeces at all, immediate treatment is necessary. It is not an option to wait and see, as this will aggravate the situation, lead to further pain and dehydration, and become harder to treat when one finally gives appropriate care (see what to do in the box on p. 125).
> - Does the rabbit have a stinky bottom? Does it have moist stools that get caught in the fur? It is important to determine if the condition is diet-related or a symptom of other diseases.
> - Does the rabbit leave dark, mushy, smelly, grape-like faeces around the house? The rabbit should normally eat these caecotrophs directly from the anus and if the rabbit does not eat these it could be a sign of a diet too rich in protein and/or too low in fibre. An improved diet should be offered.
> - If the consistency of the caecotrophs is more like a paste, an improved diet should be offered.
> - Everyone should be aware of how a healthy stomach feels. Digestive disorders may lead to abdominal distension and one should be able to notice this.
> - What is the rabbit's normal weight? Deviation must be detected. This is why regular weight checks are helpful.
> - Does the rabbit have a watery eye?
> - Does the rabbit show further symptoms of pain? (Read about how to detect this on p. 93).

> **Help, my rabbit is not eating and I see no droppings. What shall I do?**
>
> Rabbits are dependent on a constant food intake and become easily dehydrated when going off their food and water.
>
> Rabbits that have not eaten or passed faeces during the last 24 hours must be taken to the veterinarian as an emergency. However, you can provide first aid earlier by:
>
> - giving oral fluid;
> - giving nutritional support (see below); and
> - contact your veterinarian and follow the instructions.
>
> Read about first aid in Chapter 5.

Digestive System

The rabbit is a small prey animal and its anatomy reflects this. The skeleton is almost bird-like, with thin and brittle, light, yet strong cortices and it comprises only 7–8% of the animal's body weight, which is significantly less than most other mammals. That of the cat, for example, is 12–13%. This, together with the relatively enormous hind legs and lumbar muscles gives the rabbit its stunning acceleration. From a seemingly relaxed position it can shoot forwards like a cannon ball, change direction quickly and get away from the predator. Their rapid gut transit time means that the rabbit can consume and rapidly excrete large amounts of fibre. The digestive strategy of caecotrophy further minimizes gut weight by utilizing two passes through the gastrointestinal tract. This reduces the rabbit's body weight, which is a good survival strategy when you are going to escape from foxes and other enemies.

Larger prey animals are able to carry complex gastrointestinal tracts. Cows, for example, rely on a heavy foregut, while the horse relies on their large hindgut. However, rabbits are too small for this to be fully successful, and so they have evolved to minimize their weight relative to muscle mass, subject to the problem.

Traditionally, domesticated rabbits have been fed highly inappropriate food items such as grass clippings, bread, carrots, fruits, cabbage, potatoes, breakfast cereals and other leftovers from the household. Mixes for rodents and grain-based diets have also been given, often based on diets offered to rabbits in production systems, which are designed to induce rapid growth before the rabbit becomes someone's dinner or coat. However, this feed is not based on rabbits having a long and healthy life and should therefore be avoided. Rabbits are obligate herbivores and need the high amount of fibre provided in hay and grass to maintain a balanced microflora in the digestive tract and persistent dental health. The digestive system is designed to digest large amounts of hay and grass. All the processes involved are interdependent and everything starts with the first chew.

Teeth

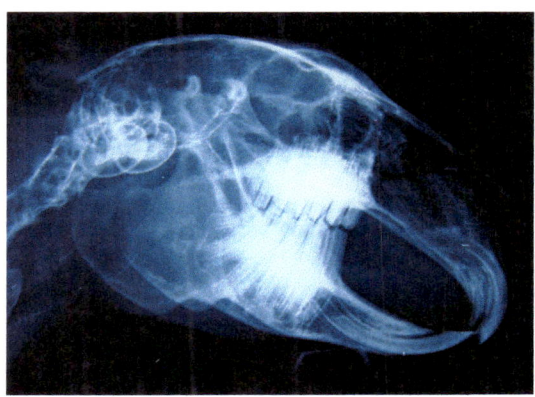

X-ray showing the teeth of a healthy rabbit. Photo courtesy of Cordelia Bracht, Norway

Rabbits have a highly specialized dentition. Along with other lagomorphs they are distinguished from the rodents by the presence of a second, smaller, pair of upper (maxillary) incisor teeth, the so-called peg teeth, in addition to the first, main pair. The lower jaw (mandible) possesses a single pair of incisor teeth. Between the incisor teeth and the premolars lies a space, or diastema, and no canines are present. In addition to the front teeth the rabbit has several cheek teeth (premolars and molars), six on each side in the upper jaw and five on each side in the lower jaw. An adult rabbit has 28 teeth in total. Rabbit's teeth are termed 'aradicular', meaning that there is no true root, but instead there is a soft germinal tissue or pulp that forms new dental cells so that the teeth will grow continually throughout life.

Acquired dental disease is nearly always related to poor nutrition. Rabbits are herbivores and dependent on a diet based on hays or fresh grass to maintain a normal gut flora and viable dental health. Grass plants have evolved concurrently with the herbivores' adaptation to eat them, and most of the plants in the grass family have microscopic silicates in their leaves, which are highly abrasive and wear down the teeth of the animal eating them. In addition, the grasses' shape and plant fibre content (cellulose and the digestible fibre) requires a figure of eight grinding jaw movement for ingesting the plant, with both side- and back-and-forth-movements in the lower jaw relative to the upper jaw. This wear pattern is instrumental in maintaining the teeth shape and occlusion and avoiding the development of acquired dental disease. Eating insufficient amounts of grass and hay is a major factor in development of malocclusion, which in turn can lead to serious and painful dental diseases.

Dental wear is affected by the abrasive nature of the diet, as explained above. Rabbits can also be seen grinding their teeth when at rest, which will also lead to attrition, or occlusal wear of tooth against tooth.

It is not that long ago since dental disorders in rabbits were only believed to be congenital, and even though this occurs, it is evident that most of the dental issues seen in companion rabbits are the direct result of an incorrect diet. Rabbits with unrestricted access to hay and grass and just a limited amount of supplementation seem to maintain the shape and occlusion of their teeth.[6]

Dental problems

Some breeds seem to have more issues with their teeth than others, especially the ones bred to be short nosed. Due to the facial shape of short-nosed breeds such as the Netherland dwarf, such rabbits are more prone to congenital and acquired problems with their teeth than rabbits with longer faces. However, everyone with rabbits in their household must be aware of the clinical signs that may indicate serious dental diseases. If one detects symptoms, a veterinarian must be consulted, as dental illness is a serious welfare concern and requires specialist treatment.

Symptoms of dental diseases

- Decrease in food consumption or change in food preferences.
- Anorexia.
- Overflow of tears from the eyes (epiphora).
- Weight loss.
- Drooling.
- Bad breath.
- Pain.
- Lack of grooming.
- Uneaten caecotrophs.
- Digestive disorders.

Other problems arising from dental disorders

- **Epiphora**: an overflow of tears onto the face, due to elongated incisor roots putting pressure on the nasolacrimal (tear drainage) duct, or because the cheek teeth push against the eyeball from behind. Elongated incisor roots with resultant inflammation is the most common cause of runny eyes.
- **Dacryocystitis**: secondary bacterial infection of the tear ducts, resulting in purulent discharge from the eyes and/or nose.
- **Wounds** on the tongue and inside of the cheek because of dental spurs and elongated crowns.
- **Infections** in the jawbone, which may progress to large facial abscesses.
- **Digestive disorders**, including caecotroph accumulation, diarrhoea and increased risk of gastrointestinal foreign bodies. Further preferential selection of more easily eaten foods, and avoidance of hay and grass, resulting in an increased incidence of digestive disturbances and potential obesity.
- **Myiasis** (flystrike).

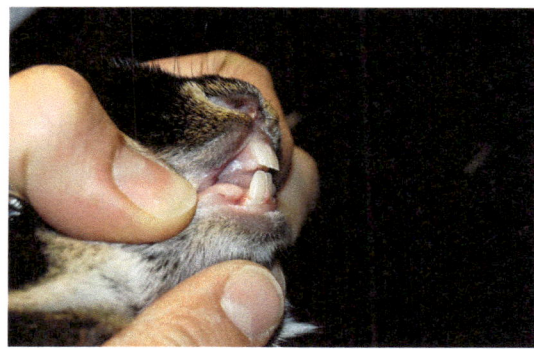

Melis has perfect dentition. Photo courtesy of Marit Emilie Buseth, Norway

The need for a balanced diet

The most important and best strategy one can implement for preventing dental diseases in rabbits is to provide an appropriately abrasive and nutritionally balanced diet. Suboptimal diets are unfortunately still recommended and fed throughout the world. Diet is so crucial for dental health that the knowledge must be provided for both veterinarians and owners.

There are two main competing theories to explain the high incidence of acquired dental disease in rabbits. One suggests that this is due to a simple lack of abrasion of cheek teeth, due to eating food that does not require a correct grinding action and which is insufficiently abrasive. The other suggests that the problem is due to a change in position of the cheek teeth within the bones of the skull due to metabolic bone disease. However, both of these causes are addressed by correcting the diet to approximate that of the wild rabbit.

The most common problem with rabbit nutrition is the provision of too much supplementary 'dry' food, compared to forage and fresh vegetation. This does not only affect the digestive tract, but also the rabbit's jaw movement. The masticatory movements while eating pellets are not as effective in wearing down dental enamel as the sideways grinding they use when eating grass.

Spending much of their waking state grazing, wild rabbits will spend more time chewing. This consequently wears their teeth both at a faster rate and in a pattern more appropriate to dental wear and occlusal forces than a pet rabbit provided a bowl of concentrated mix. Both the mastication itself and the time spent on chewing is of great importance for the dental wear, and the rabbit should therefore be offered a diet based on grass and hay, with some dark green leafy vegetation, and a limited amount of fibrous pellets for mineral and vitamin supplementation.

In addition to wearing of the teeth, correct nutrition is also crucial for providing necessary nutrients of importance for dental health. The constant attrition and growth of the teeth demands a balanced supply of calcium, phosphate and vitamin D to prevent demineralization of the surrounding bone and consequent dental malocclusion. During their growing period, young rabbits will also have an even greater need for a correctly balanced diet, as the development of their bone, skull and skeleton demands an appropriate level of calcium. Problems connected to calcium metabolism are especially an issue for rabbits with limited access to sunlight, as the formation of dentine and enamel is dependent on the rabbit's calcium level, which, at low dietary levels, requires vitamin D to be absorbed. Vitamin D, where not present or not adequately absorbed from the diet, needs to be produced by the action of sunlight on skin. Rabbit vitamin D metabolism is controversial, but rabbits exposed to low levels of sunlight have been shown to produce insufficient vitamin D.[7]

Grass, hay and forage-based pellets ensure high levels of fibre and adequate levels of calcium. Unsuitable but unfortunately widely used diets based on grains, fruit and seeds have a calcium:phosphorous ratio imbalance when selectively fed.

A balanced and abrasive diet is thus so crucial for the development and maintenance of the rabbit's dental health that it can be seen as a prophylactic or preventive precaution (see recommended diet in illustration on p. 117).

Jokk and Amos eat large amounts of hay. Photo courtesy of Katarina Vallbo, Sweden

Other reasons for development of dental diseases

While nutrition and dental wear seem to be the critical factors for dental health in rabbits, there are other possible explanations for diseases as well. Other reasons why the teeth do not wear evenly include:

- *Congenital defects or genetic predisposition.* Short-nosed breeds may be over-represented with malocclusion than the original more long-faced rabbit breeds, but it is most important to avoid breeding of animals with malocclusion in their family. Netherland dwarfs, for example, are a common breed, and are over-represented in terms of congenital mandibular prognathism, or overlong lower jaw relative to the upper jaw. Lionheads are an increasingly common breed, selected for their appealing fluffy fur around the neck and face. They are predominantly derived from Netherland dwarf stock and share the above problem. Selection for one characteristic often limits the genetic variation within a population, and allows other characteristics, which may be problematic, such as dental malocclusion, to appear in greater prevalence. Alternatively, the feature selected for, for example a short, round, appealing face, may itself be the problem.
- *Trauma* such as fractures after falling or being dropped, or head trauma after an attack by a predator. Fortunately, road accidents and falls from a height, common causes of facial fractures in the cat, are uncommon in the rabbit, and leaping out of the grasp of handlers is probably the most common injury to befall the domestic rabbit. Fractures of teeth are also common after trimming of the incisors using nail clippers, which unfortunately was a widely used treatment for malocclusion of the front teeth in the past. For current recommendations on the correct treatment of incisor elongation and malocclusion, see below.
- *Tumours* or *infection* of the soft tissue or bone at the apex of the developing teeth may distort the germinal tissue and lead to malocclusion.

Dental disease in rabbits can also be caused by several factors that affect each other. For example, malocclusion of the molars may cause overgrowth of the incisors, and genetic disposition may make some rabbits more vulnerable towards nutritional issues than others.

Incisors

In healthy rabbits, the incisors wear against each other while they are chewing. Because of their shape and occlusion, the front teeth are worn to a cutting edge, sharpening themselves as they slice. Dental enamel is only present on the labial side of the front teeth, creating the classic chisel shape. This, in conjunction with the animal's tiny overbite, explains why the incisors are as sharp and effective as scissors, slicing through vegetation.

A second set of small teeth, the 'peg teeth', is located behind the upper incisors and prevents the lower front teeth impinging the upper gum. The peg teeth are also known as the 'lagomorph teeth', and distinguish rabbits and hares from rodents.

Malocclusion and other dental disorders of incisors

The upper and lower incisors grow 10 and 12 cm every year, respectively. When they fail to meet correctly, the teeth will not wear evenly against one another and will therefore grow abnormally.

The condition is obvious once the lips are parted and the teeth examined, but it is important to know what to look for and to examine the rabbit's teeth regularly. Rabbits are excellent at hiding pain and weakness, and malocclusion and resulting elongation can develop to a significant degree before the owner notices symptoms of the condition.

Few clinical signs are seen and the rabbit may even eat with incisors that painfully curl outwards.

Malocclusion of the incisors is more commonly due to genetic factors, than in the case of the cheek teeth. Inherited malocclusion is often seen at an early age (typically before 6 months of age), and one should never breed with rabbits that have this recessive trait in the family.

Even though many of the dental issues of the incisors are congenital, malocclusion of the front teeth can also be part of acquired dental disorders as explained above, with root elongation and deterioration of the teeth.

Abscesses, tumours, infections in tooth, root and bone, traumatic injuries after accidents, such as being dropped, or damage from gnawing of the bars will also affect the dental position of the front teeth.

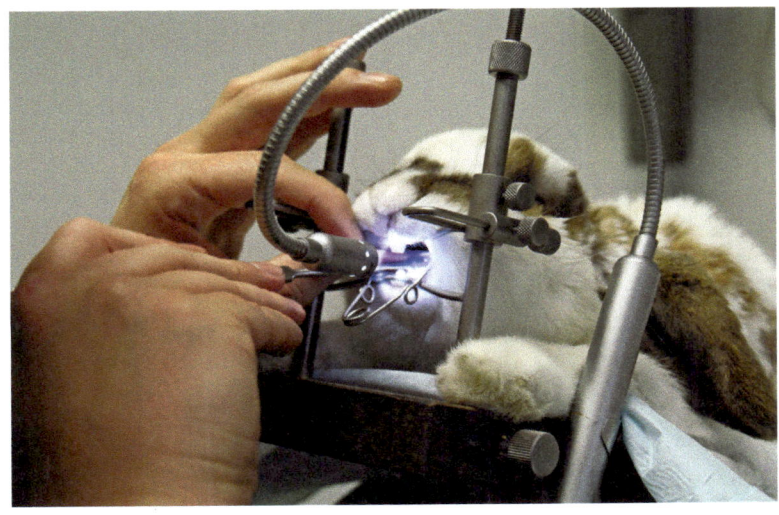

Dentistry. Photo courtesy of Smådyrakutten, http://www.smadyrakutten.no

These damages may also lead to a blockage of the nasolacrimal duct since they are located right beside the long tooth roots. A watery eye is therefore one of the first sign of dental disease, and the reason why a full dental examination should be carried out in the case of epiphora.

Examination and treatment of the incisors

It is easy to visually examine the incisors, simply by retracting the lips. Viewing them from the front and sides is necessary to reveal subtle abnormalities.

However, a thorough examination of the entire oral cavity requires anaesthesia and radiography.

> Incisors should never be clipped or trimmed with nail clippers, as this leads to unpredictable shattering of the teeth, often below the gum line, and loosening of the teeth in position. Exposure of the sterile pulp cavity and bruising to the apex of the tooth can occur. Both of these factors may contribute to pulp infection and future abscess formation and distortion of the normal anatomy of the tooth, resulting in worsening dental disease. It is also important to stress the fact that one should never let anyone but a skilled veterinarian treat rabbit's teeth.

Treatment of maloccluding incisors will often include removal of the front teeth, especially in young rabbits suffering from congenital incisor malocclusion. The rabbit will adapt to such a condition and easily eat hay, pellets, sliced vegetables and leafy veg with the tongue and the cheek teeth, but may need some assistance with future grooming.

Cheek teeth (premolars and molars)

Rabbits' teeth are arranged to wear against each other while eating abrasive food. The jaw movements during grinding of hay and grass are crucial for keeping the cheek teeth in shape. Like the incisors, their dental health is dependent on continual growth and attrition.

Herbivores have a gap between the incisors and premolars. This space between the teeth is perfect for inserting a syringe when giving either supportive nutrition or liquid medications orally (see pp. 94–96 for advice on giving medicine).

Malocclusion and other dental disorders of the cheek teeth

The molars grow about 3 mm per month, and when they fail to meet correctly, the resulting lack of dental wear across the whole occlusal surface of the tooth will result in spurs. Spurs range between anything from the whole tooth to a thin sliver of enamel, angling sideways or backwards. These may be razor sharp and this traumatic occlusion is extremely painful for the rabbit. Spurs from the lower cheek teeth typically grow into the tongue, whilst spurs from the upper cheek teeth will grow into the cheek, both resulting in ulceration. In the latter stages of dental disease, the malocclusion of the teeth within the jaw can lead to the very backmost lower two molars angling out towards the cheek, or backwards into the gum. To prevent further pain while chewing, the rabbit will naturally stop eating in the later stages of malocclusion. It is important to be aware of the fact that they still may eat food that does not involve the sideways jaw movement used while grinding hay. If a rabbit stops eating hay while it still eats vegetables, the cheek teeth should be examined. In most cases this requires sedation or anaesthesia and excellent lighting.

Malocclusion of the cheek teeth is usually part of an acquired dental disorder where the progression can be graded, and includes both root elongation and deterioration in tooth quality. A progressive and moderately predictable pattern of dental disease ensues. The teeth lengthen, both within the mouth and under the gumline. They curve and move farther apart, allowing food material to get stuck between them. They stop meeting evenly, leading to spur formation as well as irregularities, with some teeth being much longer than the others. The enamel quality deteriorates, making them less effective at grinding hay and leading to them breaking, which can be painful as well as leaving bits of tooth to act as a focus for infection. The roots perforate the bone of the jaws, leading potentially to bone infection and abscesses. If the rabbit survives all of these stages, ultimately the teeth stop growing.[8]

Examination and treatment of cheek teeth

A veterinarian can examine and evaluate the cheek teeth in the conscious rabbit by using an otoscope or vaginal speculum, inserted gently into the mouth. However, a thorough examination of the oral cavity requires anaesthesia. Abnormalities of the crowns can be seen by visual examination, but radiography or other imaging modalities such as CT (computerized tomography, a 3-dimensional computer-rendered series of X-ray 'slices') of the skull will give information on the roots and potential abscesses lying beneath the gum line and in the bone.

The main problem with diagnosing dental conditions is that all the cheek teeth are deep within the mouth and cannot easily be seen. Even the incisors can only be seen above the gum line, and there can be significant pathology present below the gums. Radiographs and CT help to image the whole head, and good visual examination of the back of the mouth, under sedation, or ideally anaesthesia, is vital. Magnification and illumination of this dark and relatively inaccessible place is important, and one way to achieve this is with the use of an endoscope. An endoscope is a fibre optic tube with a camera or viewer and a bright light source on one end, enabling very close-up inspection of even the most distant parts of the inside of the mouth. It is invaluable for looking between teeth, a view that is almost impossible to achieve in any other way. Endoscopes are expensive, and require practice and training to use well.

It is important to note that by the time significant malocclusion has developed within the mouth, and under the gum line, the teeth are not meeting their opposing numbers correctly and so regular treatment is necessary. Correcting the diet to include a much higher proportion of highly fibrous foods is vital, where possible, but the elongation of the 'reserve crown', that part of the tooth lying under the gum and within bone, has often led to it lying in close proximity to the mandibular alveolar nerve, leading to pain when grinding hard foods. This explains why many rabbits, even before problems are otherwise noted, appear to avoid hay in preference to pelleted foods. If the rabbit can be fed more fibrous foods, this may increase the wearing of the teeth and extend the interval between treatments to shorten them and remove spurs. However, the underlying anatomy has been altered, and these teeth will continue to develop spurs until they cease to grow.

Cessation of growth may occur relatively rapidly from this point on, as the bone and germinal pulp tissue of the rabbit's teeth is now diseased and unable to continue producing enamel of sufficient quality, or indeed at all, after a while. At this point, the rabbit may become more comfortable due to the lack of sharp spur development.

Until then, regular trimming of the sharp spurs and shortening of grossly elongated cheek teeth is necessary to prevent them from causing pain and damaging the soft tissues inside the mouth.

There is controversy as to the precise interval between dental treatments and the amount of tooth that should be removed at each procedure, but intervals of between 1 and 6 months are typical, with most rabbits needing 2–3 monthly cheek tooth shortening until growth ceases.

Other options for treatment include the removal of the pulp tissue within the enamel to prevent further growth of the tooth, and removal of the entire tooth itself. However, these, especially the latter, are not first-line treatments and, as dental problems rarely affect single teeth, are unlikely to resolve the entire dental problem for the rabbit.

Dental disease results in severe pain following damage to soft tissues within the mouth. Prompt removal of spurs and appropriate pain relief is mandatory to maintain good welfare in affected rabbits.

Abscesses

One common consequence of elongated crowns, fractured teeth or other dental diseases is abscess formation in and around the mouth.

Removal of such abscesses is considerably more challenging than in other species, as rabbits have particularly thick purulent material (pus), surrounded by a thick abscess capsule. As a result, antibiotic treatment rarely penetrates the abscess and is generally ineffective if used as a sole treatment. Pus does not drain readily from incisions made into the abscess, and so lancing is generally ineffective. It is helpful to obtain samples of the capsule to perform bacterial culture testing, to know which antibiotic to use in treatment.

Abscesses often invade, or arise from, bone, and this resulting bone infection, or osteomyelitis, is also extremely difficult to counter using antibiotics alone. Long courses of treatment are needed, and typically, surgical exploration and removal (debridement) of the affected bone, tooth and other involved tissues is required to effect a cure.

Such surgeries may result in an open wound left to heal by granulation, with frequent local treatment to remove discharge and avoid further infection; or the incisions may be closed, often leaving local treatment at the site, in the form of bone cement, impregnated with appropriate antibiotics.

Abscesses, whilst perhaps not as acutely painful as in some other species, are doubtless a cause of pain and discomfort. Rabbits are, of course, very good at masking this pain, but given the potential for welfare problems, dental disease should be thoroughly investigated in such animals and appropriate treatment given to relieve suffering.

Treatment options may include extremely aggressive surgical procedures, including enucleation (removal of the eye), and removal of significant amounts of bone and numbers of teeth. The long term welfare of the rabbit is paramount in deciding upon an appropriate treatment option.

Digestive Physiology

To share the apartment with three rabbits is like living with a horse. The consumption of hay is impressive. However, I doubt that a horse owner is just as happy to see horse muck as a rabbit owner is to see firm and fibrous droppings after their rabbits. As the amount and type of faeces produced are an indicator of gut function, one must always be aware of how they look and smell, as well as the quantity of droppings. Abnormalities can be a sign that something is wrong, either with the diet or the rabbit's health.

To understand why a balanced and fibrous diet is required, one should be familiar with the main points of the digestive system.

The specialized digestive structure is dependent on constant movement. A fibrous diet will encourage motility and ensure the rabbit's high rate of metabolism and correspondingly effective digestion. The large intake of fibre pushes food that is eaten through the gastrointestinal (GI) tract, ensuring that undigested food does not stay there for long. This means that rabbits can consume large amounts of grass and still have a light body weight.

In addition, rabbits have a huge population of bacteria in the GI tract. The bacterial flora and GI motility are dependent on each other and crucial for the rabbit's health.

Enzymes and acid in the stomach will initiate digestion of nutrients before they pass on to the small intestine. Rabbits do not break down fibre completely and the indigestible and digestible fibre components are separated and sent in opposite directions. The rabbit colon gets rid of the indigestible large particles as soon as possible, and this helps to push out other indigestible material the rabbit may have eaten, such as fur, carpet or plastic, before it ends up as the usual type of rabbit poo (brown and dry droppings).

The digestible fibre, however, is sent directly into the caecum, which is the rabbit's largest organ. Bacteria, minerals and vitamins are added, so that fermentation can take place before the mixture is transported back to the colon, where small balls are covered by mucus before they pass as caecotrophs. These mushy grape-like balls are further eaten directly from the anus and are rich in protein and essential nutrients, something that helps the rabbits to make use of a nutrient-poor and fibrous diet.

Faecal pellets

These are fibrous, brown droppings that mainly consist of undigested hay and grass. They should be round and firm with no liquid. The odourless droppings should be dry and easy to pulverize. If the droppings are poorly formed, small and dark, it is most likely a sign that the rabbit needs to eat more hay. Small, dry droppings strung together like a necklace with fur are a sign of reduced gut motility and too much fur in the system. A fibrous diet, water and grooming of the coat is often a quick fix.

Caecotrophs

These take the form of dark, mushy, smelly, grape-like balls pressed together. The rabbit normally eats them directly from the anus. If the rabbit does not eat the caecotrophs, it could be a sign of a diet too rich in protein. If the consistency is more like a paste, an improved diet should be offered. If the conditions persist a veterinarian should be consulted.

Droppings strung together like a necklace with fur. Photo courtesy of Marit Emilie Buseth, Norway

This process is called coprophagy and makes it possible for rabbits to have such an effective and rapid digestion and still obtain the necessary amount of nutrients.[9]

Compared to the normal droppings, caecotrophs contain twice the amount of protein and half the amount of fibre. When the caecotrophs are eaten again, previously undigested sustenance will be absorbed and the rabbit will therefore take nourishment from the food twice. In addition, the period that the digestible fibre spends in the caecum allows it to be broken down by caecal bacteria, and fatty acids are produced, which are one of the main sources of energy for the rabbit. The caecotrophs will remain intact until the mucus dissolves after 6 hours and the rabbit can finally utilize the nutrients. Wild rabbits graze most at dawn and dusk, and coprophagy makes it possible to remain in the burrow during the night, hiding from predators.

The digestive tract is relatively long and can make up a total of 20% of the body weight.

To make room for this proportionately large GI system, the abdominal cavity is significantly larger than in similar sized animals. The stomach itself is moderately large and thin walled, which makes the elasticity poor. A bloated stomach filled with gas, due to the effect of bacterial fermentation on food material, is therefore extremely painful for the rabbit. It is also worth noting that a sphincter by the oesophagus ensures that the rabbit is unable to vomit.

A fibrous low-carb diet

As a consequence of a rapid digestion, starch is not completely broken down in the rabbit's small intestine. The partial chemical digestion leads to an accumulation of microorganisms in the GI tract, and a carbohydrate overload will thus lead to an expansive bacterial growth with consequent imbalance in the microbial population.

Easily digestible carbohydrates found in sugar and starch should therefore be avoided. High fibre levels drive optimal gut motility due to increased motilin (a hormone that stimulates GI movement) production by the small intestine, and also provide the optimum environment for caecal microflora. A low fibre diet can, therefore, result in gut hypomotility, causing food to remain in the GI system for a long time.

It is a fibrous diet based on grass that ensures ingesta is pushed through the gastrointestinal tract, preventing fur and other indigestible material and food from staying put and causing disturbance in the microbial population.

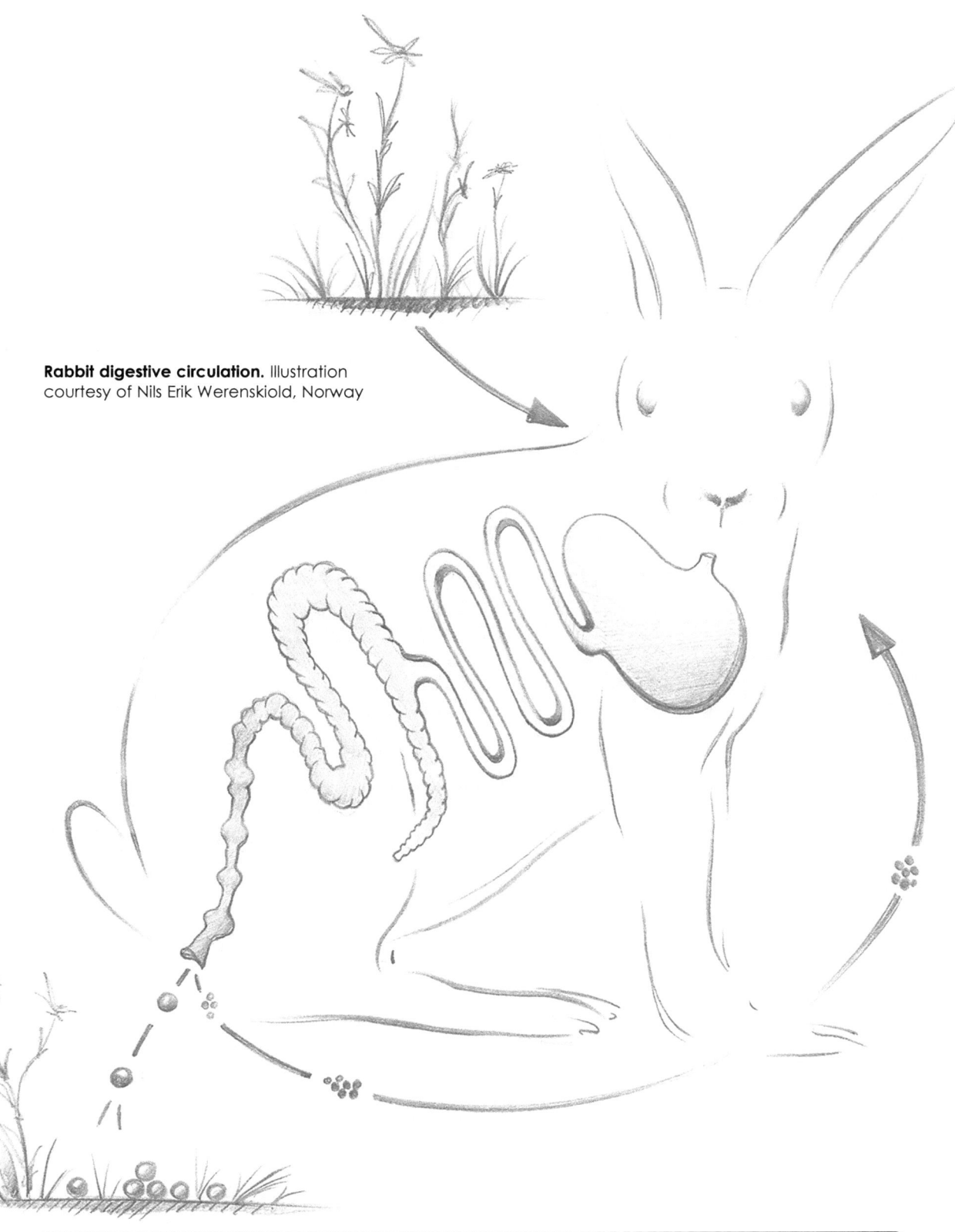

Rabbit digestive circulation. Illustration courtesy of Nils Erik Werenskiold, Norway

Sandor eats grass and hay to keep healthy. Photo courtesy of Katarina Vallbo, Sweden

Rabbits should therefore be on a fibrous low-carb diet. Muesli and mixes are unfortunately high in carbs, in the form of grains and seeds, and should not be offered (see illustration on p. 117 for diet).

Digestive disorders

Gastrointestinal stasis

Motility disorders of the GI tract are extremely common and are a potentially life-threatening syndrome in rabbits. In GI stasis, the muscular contractions are slowed and will in some cases cease to function. This will result in dehydration and impaction of the normal stomach contents, and the rabbit will produce smaller and firmer faecal pellets, ultimately progressing to zero. The reduced gut motility thus leads to dehydration of gut content, which in turn decreases motility further.

Reduced GI motility reduces glucose absorption and the supply of nutrients and fluids to the caecal microflora. Alteration in caecal fermentation patterns can then lead to changes in caecal pH and short chain VFA (volatile fatty acids, a source of energy for the rabbit) production. As the balance of caecal microflora changes there is an opportunity for the proliferation of pathogenic bacteria such as *Clostridia* spp.

The rabbit's GI system is complex and sensitive, and it can be difficult to uncover the underlying cause of the digestive problems. Unfortunately, owners often bring their sick rabbit to the veterinarian too late, and stabilization of a dehydrated and weak rabbit in severe pain must be the priority before further investigation and treatment can be started. Owners should therefore always be alert and aware of what they should look for and react to (see Chapter 5).

Intestinal obstruction

Gastrointestinal stasis was previously explained as an accumulation of fur. It is more likely, however, that fur balls (trichobezoars) are the result of reduced motility rather than the cause of it. Water is withdrawn from the intestine when the gut motility decreases, and the resulting dehydration leads to an impaction of stomach contents that will have difficulties passing through.

The rabbit is dependent on an almost constant intake of fibre, and there will always be food in the digestive system. Hair from normal grooming will also be present and will simply pass out in the faeces. However, stasis will consequently lead to longer transit time and resulting dehydration. When the content dehydrates this can lead to the formation of a ball consisting of both undigested food and fur. In the case of cats, the fur ball consists almost entirely of fur, while the rabbit's in most cases is a ball of undigested food that is held together by large amounts of mucus and hair. They will still appear as pure hairballs in X-rays or by an autopsy.

There are exceptions to the above assumptions that hair balls are the result of GI-stasis. Extreme amounts of hair, shedding, tufts of paws of long-haired rabbits, may be large enough to make a blockage in the GI-tract. Corresponding obstruction may also occur when the rabbit eats carpets, plastic, cat litter, sweetcorn, dried peas and bean seeds. Similar foreign bodies will appear to make an obstruction in the small intestine.

A true obstruction will cause tremendous pain, since gas and ingesta do not get released from the GI tract, and one must remember that rabbits can neither vomit or burp. If the digestive system stops completely, the rabbit may go into shock, become rapidly dehydrated and the body temperature may drop. Low temperature is very serious and it is important to stabilize the rabbit by keeping it warm, provide pain relief and ensure fluid therapy.

An emergency situation will rapidly lead to death if not treated immediately, which can involve surgical removal of the foreign body. If the situation is less acute, and does not involve a discrete obstruction, the vet should always try to solve the problem medically. Surgery should be the last resort, as it is an extensive and risky process. If it is proved that the rabbit has a passage so that ingesta can go through the system, one can rather provide the rabbit with curative care and medications.

It is very important to ensure that the rabbit has passage before providing motility-enhancing drugs that may increase the pressure in the intestines and stomach and aggravate the rabbit's condition.

What triggers gastrointestinal stasis

GI-stasis is usually the result of feeding a low fibre diet but may also be triggered by other factors – either obvious clinical conditions such as dental spurs, sore hocks and wounds, or conditions with no obvious clinical abnormalities. It is also worth pointing out that when stressed or in pain, rabbits release catecholamines (such as adrenaline), which act to decrease gut motility. Factors that may trigger stasis can be:

- Experience of stress, such as changes in the environment, boredom, loss of a friend, threat from a dog or cat, transport, extreme weather conditions or agitation in the group.
- A changed diet and consequently changes in the bacterial flora.
- Dehydration.
- Anorexia.
- Pain.
- Obesity.
- Lack of movement.
- Ingestion of foreign indigestible objects, such as carpet, plastic, fur.
- Adhesions.
- Toxins.

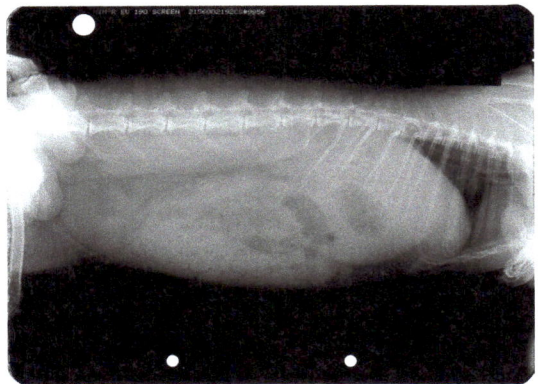

Healthy rabbit, with normal radiographic appearance of the abdomen. Photo courtesy of Tonje Haug, Norway

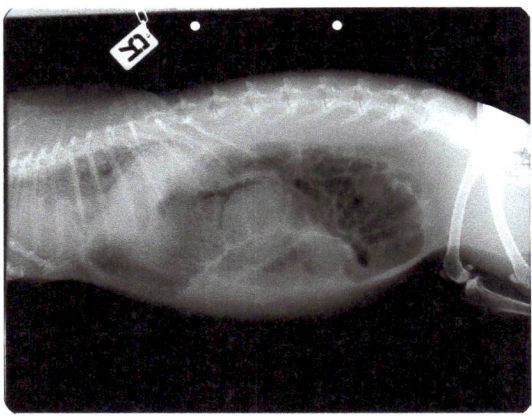

Sick rabbit, with extremely large amounts of gas filling the caecum. Photo courtesy of Tonje Haug, Norway

Difference in non-obstructive and obstructive ileus

Different causes of the rabbit's condition require different treatment, and a correct diagnosis can be crucial for survival.

The veterinarian will try to diagnose the rabbit correctly, which means that they, in addition to the clinical examination, will ask some questions to get to know the rabbit, such as:

- What is the rabbit's diet?
- How are the living conditions?
- What kind of litter is in the rabbit's box?
- Have there been some changes in the rabbit's life lately?
- Are there any changes in the rabbit's appetite and eating habits?
- When did it last pass faecal pellets? What did these look like?
- Has the rabbit changed its behaviour lately?

Together with the clinical examination, the answers will help the veterinarian to gain an understanding of condition and consequently help the rabbit in the best way possible.

- The veterinarian will observe the rabbit and look for signs of pain (see pp. 89–92).
- Is the rabbit dehydrated?
- The veterinarian will assess the general condition. Matted fur or clumping of faecal material/caecotrophs around the anus will for example indicate a lack of grooming, which may be due to obesity, musculoskeletal pain or dental disease.
- A dental examination will be performed, and this is of particular importance as dental pain is a very common cause of stasis. In addition, affected rabbits may have damp dewlaps and forelimbs due to drooling.
- Abdominal palpation will give information about pain, distension of organs, masses or an impacted stomach or caecum.

Gastrointestinal (non-obstructive) stasis	Intestinal obstruction
Gradual onset (days to weeks)	Sudden onset anorexia
Gradual decrease in faecal pellets' size and production	Sudden stop in faecal pellets
Gradual onset of depression and abdominal pain	Moderate to severe depression with abdominal pain and guarding
Mild to moderate dehydration	Shock and severe dehydration
Impacted material in stomach and caecum on radiographs	Fluid and gas proximal to site of obstruction on radiographs

> **Treatment of gastrointestinal stasis**
>
> Immediate treatment is required to prevent further pain and death. Rabbits should be hospitalized and necessary medical treatment must be carried out.
>
> - **Fluid therapy.** In severe cases intravenous fluid is necessary at 100 ml/kg/day. In mild cases rabbit owners should provide oral fluid to prevent progression of the disorder (see more about fluid therapy on p. 96).
> - **Pain relief.** Because distress in itself causes GI stasis, analgesia should always be offered when pain is suspected. The doses of drugs for rabbits may be quite significantly different to those used in other, more familiar species, such as dogs and cats. Veterinary surgeons unfamiliar with rabbits may need to consult more specialist textbooks and dosage guides, such as those listed in Chapter 5. In particular, the doses of pain killers are often much higher in rabbits, and doses suitable for dogs and cats may be so low as to have no useful effect.
> - **Assisted feeding consisting of fibre and probiotics.** Special diets for syringe feeding, e.g. Critical Care from Oxbow Animal Health, Recovery from Supreme, or Emeraid (see p. 94).
> - **Motility drugs.**
> - **Gas reduction drugs** such as those designed to treat human infant colic are often recommended. It is uncertain whether they help, as in humans they work by allowing the gas to collect into a single large bubble that can be 'burped' out, and rabbits are incapable of emptying stomach gasses this way. They are, however, not absorbed by the body, and are unlikely to do any harm at all.
> - **Exercise and movement.**

- Abdominal auscultation will be performed so that the veterinarian can listen for presence of gut sounds.

Abdominal radiographs are important to see whether there is bloating or fields of gas present in the GI system. Radiographic changes in gut stasis include impacted material in the stomach and caecum and a possible 'halo' of gas around them. Gaseous distension develops as stasis continues. In intestinal obstruction, however, fluid and gas may be seen in front of the obstruction and bubbles of gas are seen in the stomach.

Notes

[1] Clauss, M. (2012) Clinical technique: feeding hay to rabbits and rodents. *Journal of Exotic Pet Medicine* 21, 80–86.

[2] The Animal Welfare Act (2006) Available at: http://archive.defra.gov.uk/foodfarm/farmanimal/welfare/act/documents/aw-act-2006-memo-101220.pdf (accessed 25 February 2013).

[3] Frantz, R., Kreuzer, M., Hummel, J., Hatt, J.M. and Clauss, M. (2011) Differences in feeding selectivity, digesta retention, digestion and gut fill between rabbits (*Oryctolagus cuniculus*) and guinea pigs (*Cava porcellus*). Available at: http://www.zora.uzh.ch/49675/5/RF_RabbitsGPigsPhysio_revision.pdf (accessed 20 December 2012).

[4] Francis, B. (2012) Back to nature. Wild vs domestic diet. *Rabbit in Autumn*, pp. 12–13.

[5] Tschudin, A., Clauss, M., Codron, D., Liesegang, A. and Hatt, J.M. (2011) Water intake in domestic rabbits (*Oryctolagus cuniculus*) from open dishes and nipple drinkers under different water and feeding regimes. *Journal of Animal Physiology and Animal Nutrition* 95, 499–511.

[6] Harcourt-Brown, F. (2002) Dental disease. In: *Textbook of Rabbit Medicine*. Butterworth-Heinemann, Oxford, UK, pp. 165–205.

[7] Fairham, J. and Harcourt-Brown, F. (1999) Preliminary investigation of the vitamin D status of pet rabbits. *The Veterinary Record* 145, 452–454.

[8] Harcourt-Brown, F. (2002) Dental disease. In: *Textbook of Rabbit Medicine*. Butterworth-Heinemann, Oxford, UK, pp. 165–205.

[9] Harcourt-Brown, F. (2002) Digestive disorders. In: *Textbook of Rabbit Medicine*. Butterworth-Heinemann, Oxford, UK, pp. 249–291.

Tusse and Linus. Photo courtesy of Marit Emilie Buseth, Norway

7

Neutering

Adde. Photo courtesy of Viktoria Sjöberg, Finland

Evolution requires that species adapt to various environmental conditions. To ensure the survival of different species they must evolve in accordance with the demands of their environments, and the most desirable and beneficial qualities are passed to the next generation. Adaptation to different conditions in this way ensures that the species' representatives have advantages that their predecessors did not.

Adapting to different types of predators, terrain, climate, water and nutrient availability are all decisive factors in the success and survival of a species. With the number of enemies and environmental challenges rabbits must face, their continued existence has developed largely upon their ability to reproduce.[1]

Since in the wild few juvenile rabbits survive to maturity, it is important for the species' survival that they breed quickly with an early onset of maturity and relatively large litter sizes (3–9, depending on food availability) and multiple litters born throughout the warmer months of the year.[2]

Our domesticated rabbits are just as eager and fertile. It is, however, not in their welfare interest to be allowed to multiply unchecked. Living in far more protected environments, they have no need to breed, as their wild siblings do, to maintain their population. Instead of securing the survival of the species, the rabbits' ability to multiply will, in captivity, result in increased welfare issues.

A surplus of domesticated animals leads to the unfortunate fact that there will always be more rabbits than there are homes to take care of them.

This overpopulation means that ever more individuals suffer from poor husbandry and lack of knowledge. A disturbing number of rabbits are also abandoned each year, often in the form of pregnant females or does with litters.

The rabbits' talent for proliferating has long made them a successful and popular food source, for humans and carnivorous domestic animals. The species was considered to be practical and economical to house, and during the Second World War, for example, rabbits were a life-saving supplement for starving families. With the rabbit's new role as a companion animal, which can live for many years, preferably with others of their own kind, the same ability to reproduce will result in health and behaviour-related challenges. These challenges, however, are reduced or eliminated by neutering.

Objections to or 'Unnatural' Neutering

There are many who object to the neutering of their rabbits, claiming that it is unnatural, which in isolation it is. However, the very act of keeping rabbits in captivity is in itself unnatural, and neutering will in fact prevent the rabbit from multiplying, which they do so effectively resulting in the common expression 'to breed like rabbits'. They live in an artificial environment and it is our duty to optimize their living conditions as much as possible. Neutering allows important welfare benefits, and the authors strongly recommend neutering all companion rabbits, unless there are overwhelming medical reasons that preclude anaesthesia or surgery. In this chapter we will highlight many of the health and social consequences of neutering, for both sexes.

It is also worth pointing out that many of those who have objections on neutering rabbits simultaneously recommend neutering of other species, e.g. cats, on behaviour and welfare grounds. It is likely, therefore, that objections are based on outdated notions of rabbit behaviour and anthropomorphic considerations rather than science.

Definitions

Neutering is defined as the removal of the reproductive organs, in either the male or female. Neutering is termed castration in the male and ovariohysterectomy in the female. Here, neutering is used to describe either procedure.

The un-neutered animal is defined as intact.

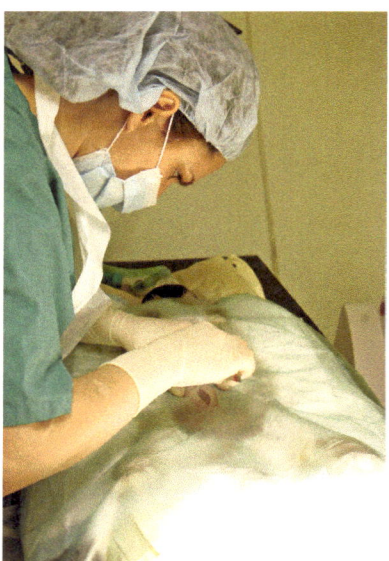

Tonje Haug neutering a female rabbit. Photo courtesy of Marit Emilie Buseth, Norway

Why should I have my rabbit neutered?

- Neutering enhances the rabbit's social life.
- Neutering is a prerequisite for letting rabbits live together, without unplanned reproduction or fighting.
- Neutering prevents pregnancy.
- Neutering will prevent cancer of the reproductive organs.
- Neutering eliminates hormonal behaviour.
- Neutering will make it easier to toilet-train rabbits.

When should my rabbit be neutered?

- Males should be neutered as soon as their testicles have descended, about 10–12 weeks of age.
- Females can be neutered safely from about 15–16 weeks of age or when their weight reaches 1 kg.

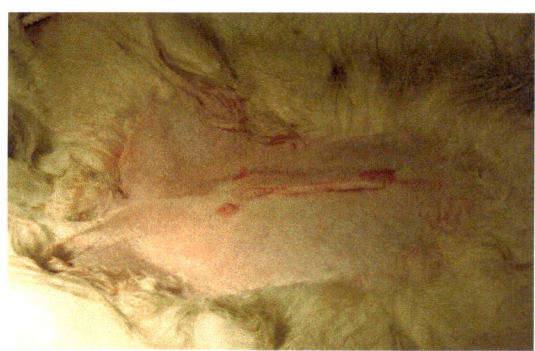

Closed up after surgery. Photo courtesy of Marit Emilie Buseth, Norway

have a physical design that allows them to give birth to a large number of offspring in a relatively short period of time. In order for this to be possible, the female has developed very active uterus tissue.

Cells of the endometrium (the tissue lining the uterus) are therefore prone to aberrant change. This may start as endometrial hyperplasia (see below) and progress to neoplasia (cancer). As well as causing local tumours within the body of the uterus, potentially leading to bleeding from the vulva, such tumours may spread locally within the abdomen, or more distantly to the brain, bone and, most commonly, the lungs.

Neutering Females

Uterine tumours

The most compelling medical reason to neuter female rabbits is their high risk of developing uterine cancer[3] (typically uterine adenocarcinoma). Rabbits

An alarming number of un-neutered female rabbits will develop uterine cancer. Uterine adenocarcinoma is the most common tumour of any type found in rabbits and occurs in up to 80% of intact does over 3 years of age.[4] Incidence does not appear to be affected by whether or not the doe has bred previously.[5]

Other uterine pathology

As a result of the fact that domestic rabbits can expect a significantly longer lifespan than their wild predecessors and relatives, they are also more likely to develop diseases of the reproductive organs, as there is a strong correlation between age and incidence, with problems more likely to develop as the rabbit ages.

Although cancer is the most widespread disease of the rabbit uterus, there are also several other diseases that are both painful and potentially fatal. These are often preventable or treatable by neutering. They include the following:

- *Pyometra*: an infected uterus full of purulent material, or pus.[6]
- *Hydrometra*: a uterus full of abnormal fluid.[7]
- *Endometrial hyperplasia*: over-proliferation and thickening of the cells lining the inside of the uterus. This may lead to abdominal pain and bleeding from the uterus into the urine.[8]
- *Uterine torsion*: twisting of the uterus, a potentially rapidly fatal condition.
- *Uterine and endometrial venous aneurysm*: rupture of blood vessels in the uterus, leading to the uterus filling with blood, with or without a bloody discharge. This may lead to life-threatening blood loss.[9]

False pregnancies/pseudopregnancy

The hormones of an intact doe can often influence them so that their body acts as if it is pregnant. This can create problems, as the rabbit will aggressively try to protect her territory. Such behaviour will affect the rabbit's relationship to both humans and other animals and is one of the main reasons why female rabbits start fighting. This is obviously stressful for the rabbit, which will go through all the activities related to being pregnant, such as nest building and milk production.

Pseudopregnancy may also have medical implications, as it shows a correlation with mammary and uterine pathology.[10] Mastitis (inflammation of the mammary glands) may also occur during lactation due to false pregnancies. Some of the affected rabbits will also experience a decreased appetite and consequently develop gastrointestinal disturbance as well.

Mammary lesions

Mammary (breast) tumours can be seen in un-neutered does from 2 years of age. Malignant tumours are not very common but can spread rapidly and be potentially fatal. Neutering will serve as a preventive intervention for the mammary tissue lesions that are noted in rabbits.

The procedure

Neutering a female rabbit is a surgical procedure, requiring entry into the abdominal cavity. This may be carried out using conventional surgery, or via a 'keyhole' or laparoscopic procedure. In the former, the uterus and ovaries are both typically removed. In the latter, generally only the ovaries are removed, making this technique more suitable for younger rabbits, in which there is little or no chance of existing precancerous changes of the uterus having started to develop. Neutering prevents the risk of developing cancer and other infections in the animal's reproductive organs. If subclinical changes are noted at surgery, then neutering is potentially curative by removing the ovaries and uterus at this stage if the disease is not too far advanced. It also allows the examination of the inside of the abdomen for local spread of neoplasia, although radiographs of the chest should be taken to examine for metastasis to the lungs.

Neutering Males

Testicular and prostatic lesions

Male rabbits also receive health benefits as a result of neutering, if not as many and as undeniable as that of female rabbits. However, hormonal-related behaviour can lead to males having fierce battles, causing damage to themselves and to other rabbits. Fighting injuries such as scrotal laceration, testicular trauma and bleeding due to wounds and cuts are not uncommon when un-neutered males are living together. Attacks are generally to the head or to the underbelly and genital area, and in severe cases the thin abdominal skin and muscle layers may be lacerated by the rabbit's powerful back legs and sharp claws, and evisceration (exposure of the gastrointestinal tract) may occur.

Neoplasia in intact males is not as common as in females, but when present it can include several tumour types. Neutering prevents both prostatic and testicular cancer.

Surgical removal of the testes also prevents orchitis, an inflammation or infection of the reproductive organs that can be brought about by a bacterial or viral infection.

Chin gland lesions

Rabbits have scent glands under the chin that they rub against objects in order to mark their ownership and presence. Intact males are especially concerned with marking of their territory. Such excessive scent-gland depositing behaviour, also known as chinning, may lead to lesions on the underside of the chin. These have been seen in a number of cases (unpublished data, Saunders *et al.*, 2014) and may require surgical excision and castration to prevent recurrence.

> Please note that rabbits can copulate through a wire mesh, and that many rabbits have produced an unexpected litter in that way. You can never let an un-neutered male and female meet each other, even under supervision, as the mating takes place in seconds.

Procedure and effects on fertility

Neutering a male rabbit involves the complete removal of both testicles. However, it is important to be aware that the male rabbit remains fertile for up to 3 weeks after the procedure, and that he should not come into contact with an intact female during this period. The level of testosterone in the blood will drop after neutering and the hormonal-related behaviour will gradually decrease. However, the sperm can still be fertile in the ducts within the spermatic cord for about 2 weeks. After 3 weeks it is dead and the rabbit can no longer multiply.

The survival of sperm in the castrated male declines more rapidly than in the vasectomized animal due to the reduced testosterone levels in the castrate, as testosterone is essential in providing a supportive environment for sperm. Periods of time of up to 6 weeks are suggested before sperm are non-viable in the vasectomized male, but this procedure does not affect behaviour and is rarely if ever recommended for rabbits.

Neutering a male rabbit is a far more simple procedure than that of a female rabbit, but both surgeries demand special knowledge of the anatomy of the rabbit, anaesthesia and recovery. In a small number of cases, usually due to delayed passage of one or both testes into the scrotum, it may be necessary to enter the abdominal cavity of the rabbit to remove them, as for the female.

> A young girl had two rabbits, a male and a female. She went to the veterinarian to get the male neutered so that they could live together. The girl and her family did not want the rabbits to multiply and asked the veterinarian whether the rabbits could meet immediately after the procedure on the male. The veterinarian ensured them that there was no problem, and the rabbits were thus introduced.
>
> The rabbits were doing fine, but after a while the female began building a nest, and a few days later she gave birth to two babies. The girl who had the rabbits was in despair, as she had not wanted to breed her animals, as she knew how hard it was to get good homes for all of them.
>
> It is unfortunate that ignorance may lead to unwanted litters. Everyone needs to know that a male can be fertile and make a female pregnant for up to 3 weeks after surgery. He cannot meet an un-neutered female until this period is over unless she is neutered herself.

When is the Right Time for Neutering?

For welfare reasons, rabbits should be neutered earlier than is often traditionally recommended. An earlier procedure has no health benefits for rabbits living alone, but for socially housed animals it may be critical to prevent both pregnancy and hormonal-related hostility.

Since living with their own species is strongly advised to maximize the rabbits' welfare, one should consider this when giving general advice on determining the time for neutering and subsequent bonding into social groups.

Even, Mandel, Harald and Melis. Photo courtesy of Marit Emilie Buseth, Norway

It is difficult to give a general recommendation for when it is appropriate to neuter a specific rabbit, as one must take into account the reason why the procedure is to be performed. The veterinarians' experience and skills with rabbit anaesthesia and surgery must of course be taken into consideration, but it is important to note that the species can be neutered earlier than has often been recommended, and that more clinics should acquire the necessary knowledge and experience to do so.

Anything between the ages of 3 and 9 months has been considered as a normal age to neuter rabbits. This is, however, not satisfactory for socially housed rabbits, as hormonal-related behaviours will occur and lead to potential problems in the group from a much earlier date.

Males can be neutered as soon as their testicles have descended, about 10–12 weeks of age. Females can be neutered safely from about 15–16 weeks of age or when their weight reaches 1 kg.

It should be noted that the procedure may be technically difficult to perform on very young rabbits, so experience and knowledge is necessary. A pre-pubertal rabbit has small reproductive organs and can therefore be more challenging to operate on. The anaesthetic risk may also increase in such tiny patients, requiring greater awareness of techniques to maintain body temperature. Conversely, older rabbits often have more fat in and around the reproductive organs, which also complicates the procedure, increasing surgical times and resulting in more perioperative haemorrhage and potentially greater risk of adhesion formation.

For medical reasons, and especially for the prevention of neoplastic changes and uterine diseases, does should at least be neutered before they are 2 years old.

The procedure can also be performed on older rabbits, but if the female rabbit is over 3 years of age, chest radiography is recommended before the surgery is performed, as above. If lung metastasis is present, this is the limiting factor in terms of life expectancy, and anaesthesia and surgery will be detrimental given the lung pathology and will carry no benefit to the rabbit.

It is a commonly and erroneously held belief that un-neutered rabbits can live in pairs as long as they have grown up together. However, siblings or early introduced individuals often begin to argue when sexual maturity occurs, which can result in severe injuries if fighting is not prevented. One can prevent hormonal-related hostility amongst rabbits living together by neutering them as soon as possible.

Many people also buy two rabbits from the same litter in the belief that they are either two males or two females. In these situations, a veterinarian or an experienced rabbit keeper should be asked to confirm the sex, since shopkeepers and inexperienced, especially accidental, breeders often have problems determining gender at 5 or 6 weeks of age, which is a typical age to sell rabbits, unfortunately. Rabbit kits should ideally be kept with their mother and siblings until they are at least 8 weeks old (see Chapter 12).

If a litter is housed together for a long time, as often is the case with rabbits in rescues, care must be taken to prevent pregnancies. Because of early sexual maturation of the males, the mother should be neutered when her kits are 12 weeks of age, or the males should be neutered or separated from their mother until this is done.

Females mature later, but when they reach 15–16 weeks of age one must ensure that they are not kept with un-neutered males.

It is the authors' opinion that males should be neutered at 11–12 weeks of age and females at 15–16 weeks of age, to prevent hormonal-related squabbling amongst males in addition to making sure no one is impregnating another rabbit.

Melis. Photo courtesy of Jens Petter Salvesen, Norway

Melis

Many people doubt the strikingly high number of potential cancer patients in un-neutered female rabbits.

The incidences are, as we have seen, medically well documented and little disputed.

The Norwegian animal charity, Dyrebeskyttelsen, and clinics we use get constant confirmation that the extent of uterine cancer is substantial. All rabbits in the rescue are neutered, and it is unfortunately more common than unusual that females older than about 3 years of age need to have tumours removed. For some of the rabbits it has unfortunately gone too far.

When Melis was found on a roundabout we at first thought she was a couple of months old. She was flimsy and weighed only 1 kg. It turned out, however, that she was underdeveloped, probably due to poor nutrition and lack of movement, and that she was in fact somewhat older than first considered.

It can be difficult to determine a rabbit's age, but a precursor to uterine cancer made it clear that she was at least 2 or 3 years old.

Fortunately, both Melis and the cancer were found in time.

Behaviour

Both male and female rabbits have a large repertoire of hormonal-related behaviour. This may involve different forms of aggression, urine spraying, spreading faecal pellets and other marking techniques, hunting and pseudopregnancy/false pregnancy.

A hormonal male rabbit can climb on to soft toys, pillows, hands and shoes, and attempt to mate with the selected object. The same rabbit will also often circle around their selected human, and many owners find this entertaining and charming.

However, a constant search for someone to mate with can be frustrating for everyone, especially for companion rabbits, who have little or no opportunity to meet a partner. A hormonal male rabbit who is constantly guarding his territory and roaming for females uses unnecessary amounts of energy that could otherwise have been used to maintain a strong immune system and other bodily functions. Some may be so keen that it even goes beyond their interest in food, and may lose weight and body condition. To avoid such stress, neutering is recommended.

False pregnancy is occasionally seen in un-neutered females. A doe suffering from pseudopregnancy will most likely occupy herself building a nest for the expected kits. She might dig and chew in order to obtain suitable materials for the nest and pluck hair from her body to insulate the nesting area further.

Rabbits in heat will have a strong desire to protect their territory, which in many cases makes them aggressive, both towards other rabbits and people around them.

After neutering such behaviour will cease.

Hormonal behaviour as mentioned above, is probably a reason why many rabbits are abandoned each year, which makes it crucial to educate owners and always recommend neutering.

Social Consequences

Rabbits are very social and prefer to live with other rabbits.[11,12] Due to the rabbit's reproductive ability and zeal, one can obviously not keep intact males and females together.

Less well known is that rabbits of the same sex also should be neutered if they are to share housing.

Aside from preventing pregnancy, neutering will also prevent hormonal-related behaviour that can be both tiresome and also lead to quarrels and

> **Rabbits' ovulation**
>
> Rabbits do not have a regular oestrus cycle. They are induced ovulators, and ovulation takes place after stimulation due to mating or even the proximity of an intact male. In the wild they are reproductively active between April and September, but when the conditions are right, as is the case for our domesticated rabbits, they breed all year round.
>
> With most mammals, ovulation takes place at regular intervals, but the rabbit has evolved to give birth to a number of kits, and can also be inseminated again directly after giving birth (at the 'post-partum' oestrus).
>
> Pseudopregnancy is the result of ovulation without fertilization, and can lead to pregnancy-like behaviour, including aggression, fur plucking and nest making.

hostility among the rabbits. One should therefore make sure to neuter both individuals in a bonded group.

Intact female rabbits living with other un-neutered female rabbits will often be good friends and can form close bonds. Periods of heat, however, will change the small group's dynamics and may lead to hostility and unrest.

Male rabbits living together with other males have a more rigid group hierarchy, but also an inherent need to guard themselves and their territory. Un-neutered male rabbits in captivity will fight and hurt each other badly as a result. Often they will attack the other's testicles and in many instances actually manage to castrate their opponent.

Neutering is thus a prerequisite for successful long-term cohabitation, regardless of gender of the animals within the pair or larger groups.

Read about social housing in laboratories in Chapter 4.

Neuter Single Rabbits?

Many wonder if it is necessary to neuter single rabbits.

For the does, the health-related consequences are so significant that the surgical procedure should be implemented regardless of the living situation.

With a single male rabbit, it is not as medically urgent that neutering is performed. However, he should be neutered if his behaviour is causing

Pia and Tjorven. Photo courtesy of Marit Emilie Buseth, Norway

Cheese and Biscuit

Cheese and Biscuit are two female rabbits living with a family. Everyone thought it was nice with rabbits in the house, but after a while Biscuit seemed upset. She chased and bit both humans and rabbit, and the family became frightened. Fortunately they decided to neuter both of them, and when they were 15 weeks old they went to a rabbit-friendly clinic.

Biscuit was delivered to the surgery the same day she was neutered. At first she went through a general health check, including an accurate measurement of weight, before she was given premedication and was allowed to rest in a dark, quiet place. When she was drowsy she was carried into the operating room, and since she was under the influence of sedation it was easy to hold the gas mask over her nose. She was sleeping within seconds.

The sleeping rabbit was turned on her back and prepared for the procedure. Gas and oxygen were supplied, and while the vet shaved Biscuit's belly, the nurse monitored the breathing and checked that everything was working as normal.

The vet removed the ovaries and uterus before the rabbit was sutured back together. This was done with a surgical suture that dissolves with time and does not need to be removed afterwards.

Cleaned and groomed, it did not take many seconds before Biscuit woke up. She was a bit confused at first, and was put in a cage to wake up properly, while wrapped in a blanket to keep warm.

Neutering seemed to be a success. I was told that Cheese and Biscuit became close companions after the surgery, and that the former somewhat angry Biscuit turned out to be endearing.

Mikke (left), one of Marikken's children, and (right) four of them. Photos courtesy of Aksel Hunstad, Norway

From 2 to 13 and the importance of neutering

It was 2 days before Christmas Eve. Most people decorated and cleaned for the holidays, while some apparently also cleared away their animals. A dirty cage with seven rabbits was abandoned at a rest stop at a road just outside town. It was snowing heavily and the dumped animals had no shelter. The scared, wet and cold rabbits were fortunately found and taken to Dyrebeskyttelsen, the Norwegian animal charity.

After arrival, it did not take long before the rabbits explored their new area, jumped on the floor and seemed pleased to have the opportunity to move about. It was apparently a family consisting of a mother, father and five children, and the male had to be separated from the others at once to prevent further propagation. He seemed surprisingly fit, while the mother was skinny and tired. Nevertheless, she took care of her offspring, which turned out to be two different litters. The two youngest babies were approximately 4–5 weeks old, while three girls were just about 2 months.

Females can become pregnant again immediately after birth and it is therefore important not to let them meet an un-neutered male. Many people are unaware of the fact that even a short meeting will result in a new litter within 30 days, and it is likely that this Christmas family was the result of ignorant owners who suddenly had far more rabbits than they expected.

The father was neutered and the whole rabbit family moved into a foster home. Max had to be kept separate from the rest of the family until the sperm and hormones calmed down, but they still had contact through the fence that divided the floor they had at their disposal. The three girls proved to be harmonious and easy to get in contact with, the boy and girl from the second litter enjoyed the life as free-range rabbits and stayed close to the peaceful Max much of the time, while Marikken, the mother, seemed to be more nervous. She was not interested in contact with humans and seemed frustrated. She didn't gain any weight and was probably exhausted after raising one litter after the other, and on the 30th day she gave birth to her third litter.

Five newborn rabbits slept in a nest of fur and hay, and now they were 13. Marikken had lots of mouths to feed and made an impressive effort, even though she was tired. Max turned out to be a calm and steady father who ensured peace and order in the group and all the children grew up to be trusting and confident rabbits.

The human in the house socialized the rabbits by sitting on the floor, grooming the ones who wanted it and just being around. She cared for them and obviously had lots of work to do with feeding and cleaning up after 13 free-range rabbits, especially before most of them were neutered and house-trained. The boys were neutered at 11 weeks of age and the girls when they were 13–14 weeks old, and there was now no need to separate the rabbits. When Marikken and her children at last were neutered and finally house-trained it was time to move outside. Spring had arrived, and enclosures and houses in the garden were waiting.

Despite the sad start, it has been a pleasure to follow the rabbit family. It has been interesting to observe and learn from their cohabitation, to see how caring the father is, experience the joy they have of each other, and watch Marikken chasing away intruders or rabbits visiting the neighbouring run. At the time of writing seven of the children are adopted, while Max, Marikken and four of their children seem to have settled down in their foster home (see the family today on p. 199).

stress, injury or disease to either himself or humans within the house. Some intact rabbits, especially males, will emit a characteristic odour. This will also disappear with neutering.

Due to the species' social needs, a single rabbit should be allowed to live indoors, together with the rest of the family. If life indoors is not desirable or possible, it is strongly recommended that the rabbit living outside gets a partner.

Fat and Lazy

Many are concerned that their rabbit will change their personality after neutering. It will not. However, what will cease is the hormonal-related behaviour. If this has been sufficiently strenuous for the rabbit, you will notice that it is somewhat calmer now that it no longer needs to hunt around for imaginary partners.

Basal metabolic rate will decline slightly after neutering, and so diet is particularly important in the neutered rabbit, but it is not neutering that makes a rabbit passive and sluggish, but an inactive life and excessive calorific intake relative to expenditure.

Notes

[1] Von Holst, D. et al. (2001) Social rank, fecundity and lifetime reproductive success in wild European rabbits (*Oryctolagus cuniculus*). *Behavioral Ecology and Sociobiology* 51, 245–254.

[2] Chapman, J.A. and Flux, J.E.C. (eds) (1991) Rabbits, Hares and Pikas: Status Survey and Conservation Action Plan. IUCN, Gland, Switzerland.

[3] Hilyer, E.V. (1994) Pet rabbits. *Veterinary Clinics of North-America: Small Animal Practice* 24, 25–69.

[4] Heatly, J.J. and Smith, A.N. (2004) Spontaneous neoplasms of lagomorphs. *Veterinary Clinics of North-America: Exotic Animal Practice* 7, 561–577.

[5] Weisbroth, S.H. (1994) Neoplastic diseases: tumors of the mammary gland. In: *The Biology of the Laboratory Rabbit*, 2nd edn. Academic Press, New York, pp. 345–347.

[6] Johnson, J.H. and Wolf, A.M. (1993) Ovarian abscesses and pyometra in domestic rabbit. *JAVMA* 5, 657–669.

[7] Morrell, J.M. (1989) Hydrometra in the rabbit. *Veterinary Record* September, 16, 325.

[8] Saito, K., Nakanishi, M. and Hasegawa, A. (2002) Uterine disorders diagnosed by ventrotomy in 47 rabbits. *Journal of Veterinary Medical Science* 64(6), 495–497.

[9] Reusch, B. (2006) Urogenital system and disorders. In: *Rabbit Medicine and Surgery*, 82nd edn. SAVA Publications, Gloucester, UK.

[10] Reusch, B. (2006) Urogenital system and disorders. In: *Rabbit Medicine and Surgery*, 82nd edn. SAVA Publications, Gloucester, UK.

[11] Held, S.D.E., Turner, R.J. and Wooton, R.J. (1995) Choices of laboratory rabbits for individual or group-housing. *Applied Animal Behaviour Science* 46(1), 81–91.

[12] Seaman, S. (2002) Laboratory rabbit housing: an investigation of the social and physical environment. Available at: http://www.ufaw.org.uk/pdf/phhsc-scholl-summary.pdf (accessed 15 February 2013).

Frigg and Balder. Photo courtesy of Aile Sandtrø

Snorre. Photo courtesy of Solveig, Norway

8

Cleanliness and Hygiene

How many rabbits can fit in a litter tray? Photo courtesy of Marit Emilie Buseth, Norway

People are often surprised when they come to my place, amazed that there is neither a bad odour of urine nor a layer of droppings covering the floor. Knowing that I have three free-range rabbits, they no doubt imagine my living conditions to be similar to some sort of barn life. However, a relatively normal apartment is modified for the rabbits, something that includes a litter tray with corresponding hay rack.

Most people know that cats use a litter tray, but few are aware of the fact that it can actually be easier to train rabbits to become house-clean. In most cases it is sufficient to neuter them and provide a correct-sized tray, although for others there is the tiresome reality of rabbits urinating on the couch and in other preferred areas. Luckily there are ways to change this!

Rabbits in a Natural Way

Wild rabbits living together in a colony will make use of a common lavatory. The toilet, which is situated outside their actual burrow, tells other rabbits that the area is occupied and provides at the same time information about its residents. Pheromones, which are chemicals or scent signals secreted by the rabbit, reveal information about an individual's social and reproductive status. Chemo-signalling through the urine and droppings enables the rabbit to communicate with passers-by of the same species, reporting both gender and ranking within the group.

Rabbits will also mark their ownership and presence by rubbing against objects with the scent glands under their chins. The dominant male's odour is usually prominent, as it is his responsibility to mark the boundaries of the property. Males of high order actually have better developed glands under both their chin and anus, and will consequently make their mark by using their glazed and fragrant droppings as well. Having a very sensitive nose, it is inevitable that a passing rabbit will get the message and hopefully move on.

This natural inclination to defecate at specific locations makes the rabbit easy to litter-train. Usually they will find one or two applicable areas and stick to these. The preferred place will normally

Edvard (on top of the box), Alfred and Emma have a trendy litter tray in their apartment. Photo courtesy of Hege Fjelde Tvedten, Norway

be in a corner or along the perimeter of the rabbit's territory. One can easily facilitate this behaviour by placing the litter tray at the selected spot. Alternatively, one can try to dictate its location, but this is best done before the habits have been established. Rabbits can be persistent creatures, and having selected a place it can be challenging to persuade them to go elsewhere. In that case it is important to have both knowledge and practical advice at hand.

Neutering and Age-related Questions

Neutering is the most effective way to house-train a rabbit. A baby rabbit can use its litter tray without mishaps until it reaches puberty when the hormones awake and it gets a sudden urge to mark its existence.

The formerly clean rabbit can almost overnight be obsessed by leaving its droppings in corners, on carpets, behind the couch, by the threshold, generally any place to which it has access.

Such a hormonally conditioned need to assert oneself will decrease after neutering. For some rabbits, this kind of marking will disappear immediately, while in most cases it will fade away within days or a couple of weeks. The degree of hormonal behaviour before the operation may have an impact on how rapidly this change will occur.

Rabbits urinate for a number of reasons. They will empty their bladder for no other reason than having the need to urinate or they will use their urine as a means of communication by squirting chosen objects or individuals. Such spraying is done with great accuracy and is not left to chance. The hormonal rabbit can spray upon walls or selected partners, all depending on what needs to be marked. Males are known to be the most active at this, but it can happen for intact does as well. Regardless of gender, the spraying will decline after neutering.

Many wonder whether it is possible to litter-train an older rabbit or a very young one. The answer is that rabbits of all ages can be clean if offered a litter tray. Older rabbits can in fact be more predictable and easier to train than their younger relatives, but some take to the concept from an early age. To continue this habit, neutering should be done as soon as possible.

Droppings and Communication

As we have seen, droppings are an important part of the communication of the species. Our domesticated rabbits will similarly mark their territories, whether they are alone in the house or not. As previously mentioned, neutering will prevent hormonally conditioned behaviour, but they will still retain the need for some social marking.

Rabbits will guard their territory. Even though most of the resource-related marking disappears in the fixed rabbit, droppings can still be spread around the floor just to signal that the area is taken. This tendency, however, is most present when groups of rabbits have shared borders.

The sudden arrival of unfamiliar rabbits in the neighbourhood can make carpet bombardment of droppings necessary. Such boundary marking is often seen when unfamiliar rabbits are living in neighbouring pens.

Similarly, a new partner might disturb previous habits. The rabbits will normally restore peace and order when they have sorted themselves out and have become well acquainted.

Due to the rabbit's personal scent, one should not sterilize the litter tray at each cleaning. Change the litter and wipe off any obvious stains, but do not clean away their identification scent entirely. The remaining odour will not bother humans, but will for the rabbits' noses make it clear who is in charge of the respective litter tray.

Litter Tray

Rabbits like to spend some time in their litter tray and if it is too small the rabbit might not want to use it. It should be sufficiently large for the whole body to fit in and litter trays for cats are normally appropriate. The bottom of a traditional cage or a simple storage box would be equally suitable, depending on how many rabbits you have housed together.

Harald, Even and Melis have plenty of room and hay in their litter tray. Photo courtesy of Marit Emilie Buseth, Norway

The manufacturers of the Rabbit Litter Box claim that it keeps urine off the rabbit's feet and bottom, stops digging in the litter, reduces litter cost and is easier to clean. Photo courtesy of Aksel Hunstad, Norway

Litter trays designed specifically for rabbits can be a good choice as well, and double-bottomed litter trays can be a solution for rabbits that kick a lot of litter out of the toilet. Simply put the litter in the bottom tray, cover with a second tray with a strainer in the bottom and let the rabbit eat hay and droppings in the top part. The urine drips through to the bottom portion, the droppings remain in the top part and can easily be picked up and dumped immediately. Make sure to use a grated surface that doesn't harm the rabbit's feet and nails.

Occasionally, the rabbit can urinate over the edge, even though it is sitting in the litter tray. This is usually accidental, as the rabbit actually jumped into the litter tray with good intentions. However, they do have a tendency to squeeze their bottom into the edge, and as they simultaneously raise their tail when urinating mishaps can occur. Luckily this problem can be fixed by simply introducing a tray with higher walls.

Socially housed rabbits will enjoy spending time together in the litter tray. Hay should always be easily accessible, and the tray will be a popular hang-out where the rabbits can chew and leave their faeces at the same time. It is therefore important that the rabbits are offered a litter tray large enough to accommodate all the rabbits in the household. As we have seen, rabbits like to urinate in the same place as others.

Disabled rabbits, due to illness, injuries or old age, may have problems getting into a normal litter tray. These should be offered a toilet with lower walls, so that the rabbit does not need to wear itself out whilst staying house-clean. If the litter tray seems inaccessible, the rabbit will of course choose an easier place to do his business. Modification of the litter tray may therefore be necessary for these rabbits.

Customized litter tray for an elderly rabbit.
Photo courtesy of Aksel Hunstad, Norway

How to remove urine stains

Pour white wine vinegar on the spot, leave for a few minutes before removing with soap and water, dependent on the material or fabric.

Litter

There are several available litters, but there are some pros and cons of the different products. A good reference here is The House Rabbit Society;[1] however, the most important factor is that owners are aware of what litters to avoid:

- Clumping litters are dangerous and must be avoided. Rabbits will always nibble some of the bedding, and then the litter will clump, almost like cement, and cause intestinal obstruction (see p. 135). The respiratory tract will also suffer, due to dust and gases.
- Shavings from pine and cedar should be avoided, as such shavings will emit gases that cause liver damage.

Recommended and safe litter:

- Organic litters can be made from, for example, pulp and recycled paper. Such litters are generally good at absorbing and cutting down on odours.
- Compressed sawdust pellets are an inexpensive and a safe alternative. They are safe to use, since the toxic phenolic is removed during production.

Hay in the litter tray

Fiona has plenty of hay in the litter tray. Photo courtesy of Marit Emilie Buseth, Norway

To increase the probability of your rabbit using its litter tray, one can simply offer them hay in it. Whether this is done by providing an attached hay manger, or by laying a bundle in a corner, it is an effective way to make your rabbit house-clean. With their tendency to relax the sphincter muscle while grazing, they will often leave droppings while they are eating. As they are dependent on eating steadily, they will spend a great deal of time in the litter tray.

With a digestion dependent on constant movement, a large amount of hay will lead to motility, a rapid metabolism and accordingly an effective digestion. Grazing large quantities of hay will result in a corresponding and healthy amount of droppings.

> If one chooses to have hay in the corner of the litter tray you will need to make sure it remains dry. Rabbits will leave their droppings and urine on the hay as well as the bedding, and an arrangement like this will subsequently require that one must replace a portion of the hay on a regular basis.

Expansion of the rabbit's area/extended area

When a rabbit moves to a new home it will naturally feel the urge to mark its new territory. Some immediately begin using the litter tray once it is placed in an approved area, while others will feel an urge to leave their mark in their new living arrangements. Initially confining the rabbit to a smaller area and then increasing this when good habits have been established is an effective solution for making these rabbits house-clean. The rabbit will most likely return to the original litter tray when given free roam of the house or the gradually expanded area.

Fences can be fixed together to create a temporary small area. Depending on what is available in different countries, one can make use of puppy play yards, exercise pens and even compost bins. (Further advice on penned off areas is given in Chapter 9.)

For the same reason, a new occupier may leave its droppings all over the place; moving around might overwhelm a rabbit. If your rabbit is living outside in a hutch and leaves its droppings everywhere when inside the house, this is why. Being a creature of habit, stability seems important for rabbits.

The tray does not need to get in the way. Reidar and his companions have an invisible solution. Photo courtesy of Marit Emilie Buseth, Norway

Balder and Frigg in their litter tray. Photo courtesy of Marit Emilie Buseth, Norway

Rabbit Habit

Rabbits have very definite opinions about how their living space should be arranged. They like to determine the location of the litter tray and one can prevent domestic dispute by simply putting the litter tray in their chosen spot. However, this is not a solution if the bed or sofa becomes the designated toilet. Such an absorbent fabric can make it tempting to leave one's mark. As good luck would have it you can change their minds.

An effective approach to this problem is to deny access to the couch for a period of time, so that the rabbit can establish more desirable habits before being given access to the furniture again. This is the same premise as explained in the previous section.

If one does not trust the rabbit and is concerned about vulnerable or exposed furniture, the sofa or bed can also be covered with an incontinence sheet intended for sufferers of urinary incontinence until the rabbit is house-trained.

Where is the bathroom?

The toilet must be readily available. Having free roam of the house, rabbits will seek out their litter tray. On the other hand, if the rabbit has a cage as a base, this should be placed on the floor, so that the rabbit can still jump in and out easily. If your rabbit wants a second bathroom it is advisable to compromise and place an extra litter tray at the selected spot.

Formerly house-clean rabbit stops using the litter tray

We have learned that rabbits are clean animals that prefer to have their toilets in particular places. When they have found a corner or another suitable location for their lavatory, simply place the litter tray at their chosen spot. Place some droppings and a sheet of paper towel with wiped up urine in the litter tray, as this will encourage further use of the toilet.

If a formerly house-clean rabbit suddenly stops using its tray, or if a new rabbit does not become litter-trained at all, the problem might be health related. A urinalysis can show different types of urinary tract infections or bladder sludge, both painful and treatable. Incontinence may also be a symptom of several other ailments, for example arthritis and *E. cuniculi*. Medical treatment and necessary adjustments of the litter tray will often solve the problem.

A mailbox filled with hay for Puma, Amalie and Nike. Photo courtesy of Ingri Bøthun Leganger, Norway

During Petter's first months he was determined to do his business in two different places. One of them was in a litter tray in the kitchen, while the other preferred spot unfortunately was in my bed. He loved to stay in bed with me, but after some frustrating months with inappropriate urination and frequent cleaning of bedding, I had to deny him access.

Petter turned out to be a quick learner. After putting him down, saying a distinct 'no' a couple of times, the point was taken and he no longer jumped up on to the bed. He understood that this area was out of bounds, and I never saw sign of him there again. I didn't use physical barriers, and Petter showed himself to be an obedient rabbit.

After 4 months with no accidents, I lifted him up on to the bed. He was immediately happy and understood that finally he was allowed to stay in bed again. From that day on, he spent lots of time there and he always went into the kitchen to use the toilet.

However, it is not usually so easy to keep rabbits away from selected sofas and beds. Therefore I recommend the use of physical barriers to avert traffic, either for periods or as a permanent solution.

Sudden changes may also be behaviour related, and also in these cases it is important to react as quickly as possible before new habits are established.

Stress can also cause a rabbit to change previously established habits. Surprise visits or a scary sound while the rabbit is in the litter tray, or changes in the rabbit's household can be a contributary factor. It might start urinating and leaving droppings just outside the litter tray, and it is then advisable to give the rabbit positive reinforcements to the litter tray. Offer the rabbit some treats while using the tray or move the litter tray to a new place it seems to prefer. Be prepared that it is not a quick fix to alter a rabbit's mind.

Litter tray in a cage

A litter tray should also be available for rabbits living in a hutch. As in normal litter training, hay should be available while sitting in the litter tray. Even if it is in a cage, the tray needs to be big enough to encourage the rabbit to spend time in it.

As rabbits like to urinate on absorbent materials, litter should only be offered in the litter tray. Having the same litter in the entire cage might confuse the rabbit, as it will appear as one big litter tray. As long as the hutch is not placed outside in a cold and humid climate, litter as bedding is not necessary.

A convenient and tidy solution. Photo courtesy of Marit Emilie Buseth, Norway

If the rabbit urinates on the plastic bottom during the adaptation period, one must clean this up immediately to keep urine off the rabbit's fur and skin. If this remains a problem I recommend the use of some kind of temporary bedding, such as carpet, towels or newspapers. However, if the rabbit is in a hutch or cage most of the time, an appropriate substrate should be offered. Any substrate should be regularly replaced and kept dry and clean, to prevent sore hocks (for more information on sore hocks see p. 102).

A run should be attached to the cage, so that the rabbit can seek the toilet when outside.

Doesn't it smell?

People frequently worry about rabbits living in the house, and they seem especially concerned about the smell. To reiterate this chapter's introduction, I would like to emphasize that the rabbit's reputation of being a stinking animal comes as a result of unhygienic living conditions. Anyone living in a confined space on a floor covered with a mixture of sawdust and excrement would have developed a bad odour.

Knowing that rabbits are very hygienic by nature and use a lavatory outside their den makes it clear that they do not like to live in their own excrement. A smelly and dirty hutch is therefore both unsanitary and leads to poor welfare for the rabbit.

Note

[1] House Rabbit Society (2011) FAQ: Litter training. Available at: http://www.rabbit.org/faq/sections/litter.html (accessed 20 February 2012).

Social eating in the litterbox. Photo courtesy of Katarina Vallbo, Sweden

Lill Strumpa. Photo courtesy of Weronica Söderlund, Sweden

9

Rabbit Housing and Conditions

We all know that rabbits jump. That it is still common to keep them hutched up is probably because few have questioned this tradition that dates back to the 1500s. However, with today's knowledge and opportunities there are no reasons to leave our companion animals in the same way that medieval monks kept their future meals.

The rabbits' home is where they will live their life. This chapter will explore and emphasize why it is insufficient for a rabbit to be kept in a hutch and will provide information about alternative living conditions that include sufficient space and proper enrichment.

Why a Hutch is Not Enough

Captivity often prevents animals from performing their natural behaviour. In the case of rabbits, restrictions on the ability to move about have serious consequences, and one should have knowledge on the species' needs and preferences when considering types of living arrangements.

Enclosures and free-range housing have in many cases replaced traditional rabbit hutches. Conventional cages have to some extent developed or been completely removed, even though unsatisfactory housing is still unfortunately a prevalent way to keep rabbits. Most people buy small and barren cages at pet stores or similar outlets, and sincerely think they are providing their pet a satisfying life. However, we currently have both expertise and opportunity to offer rabbits a better life.

Knowledge on wild rabbits' habitat, as well as observation and experiments in different adapted housing environments, gives us information that makes it easy to offer the domestic rabbit living conditions that promote their welfare. We know that rabbits need the opportunity to be active and exercise, we know that they prefer to sit on top of boxes and keep looking,[1] we know that they must have hiding places and that they have the need for a clean environment, and most of all we are aware of their need for a companion rabbit.[2]

Many rabbits are used in scientific procedures. To improve welfare for the animals involved, numerous studies on various housing conditions are performed. The physical consequences of a sedentary life and the effects of enrichment, such as extended living space, platforms, shelters, gnawing objects, hay and companions, are measured, which in turn has led to an improvement for many of the rabbits working in the laboratory. On the basis of acquired knowledge, many of these rabbits are housed in floor pens where it is suitable for them to run, dig, hide and nibble with buddies, rather than being hutched up in individual cages.

This tendency is unfortunately not currently the case for many rabbits kept as pets, and disturbingly many live alone in cages that are too small and poorly adapted. Knowing that many companion rabbits cannot expect other than endless days and nights in a hutch makes it intolerable that the available knowledge has not affected private husbandry in particular.

A thorough study in the Netherlands[4] evaluated the welfare of pet rabbits in Dutch households. The survey revealed that the average lifespan of pet rabbits is 4.2 years, while the potential lifespan is around 13 years. This tremendous distinction can be seen as a result of poor husbandry.

Research also reveals that almost half of the rabbit owners asked in the UK did not know that rabbits

© M.E. Buseth and R.A. Saunders 2015. *Rabbit Behaviour, Health and Care*
(M.E. Buseth and R.A. Saunders)

Todd, a relaxed and content rabbit with free access to the apartment. Photo courtesy of Marit Emilie Buseth, Norway

> The husbandry of rabbits in the UK is covered under the Animal Welfare Act 2006.[3] This basically states that all animals shall be provided with an environment that fulfils all their needs, and does not create or lead to welfare problems.

needed space to exercise. The PAW report 2012[5] also found that 10% of rabbits, around 150,000 at the time of survey, live in hutches that are so small that they can barely jump two hops. A further 6% of owners did not think the rabbit needed to go outside its cage and 16% of rabbits only had access to a run no bigger than the hutch. Based on this it seems crucial to inform owners about a proper environment.

The living environment of European wild rabbits can have an extent of 1600–6700 m^2, equivalent to up to 34 tennis courts. The species can run at 45 km/h, and it should be obvious that it is difficult to accommodate the same animals' needs in a hutch. Our rabbits are direct descendants of the European wild rabbits and still possess similar behaviour, desires and needs. A rabbit that has only 1 m^2 available will naturally enough experience both physical and behavioural challenges.

Gabriel has a running track in the house. Put non-slip carpets on the floor to ensure that the rabbits can move naturally. Photo courtesy of Katarina Vallbo, Sweden

Rabbits' Physical Wellbeing

Rabbits have a very light and fragile bone structure, which makes them vulnerable under caging conditions with no or few possibilities for normal activities, such as running, hopping or even just sitting up on their hind legs. Rabbits living a sedentary life

Harald, Melis and Even did not approve of this hutch. Photo courtesy of Marit Emilie Buseth, Norway

Pedro and Amos in their great run. Photo courtesy of Katarina Vallbo, Sweden

develop a poor bone structure, and those who have to spend their lives in a typical cage often suffer from thinning of their bones. In rabbits developing such osteoporosis, the brittle bone is being exposed for fracture and damage. Since the rabbit is excellent at hiding pain, it is hard to know whether your companion rabbit actually has a breakage or other abnormalities in the skeleton.

The best one can do is to prevent such problems is by providing a sufficient and facilitated living environment.

Rabbits that have been housed in standard cages for more than 6 months will be prone to injuries and fractures due to weakness in the bone structure.[6] Because of cage dimensions with no opportunity for exercise, these rabbits will lack coordination, muscles and flexibility that would have been normal for the species.

This is particularly evident in large farming systems where rabbits are being kept in small cages with no access to a run. I receive regular reports of such rabbits that suffer fractures of the spine because they have run into the wall. However, several experienced breeders I have talked to believe that this is normal and will not offer their rabbits recommended platforms or boxes they can jump up to, as they apparently think that rabbits will break their legs when jumping down. Their experience with frequent injuries clearly proves the point; the need for movement is important for maintaining a healthy rabbit. Brittle bones and fractures are also one of the most common pathologies in laboratory rabbits due to prolonged cage confinement.

Rabbits that have grown up in large and well-equipped pens show on the other hand a far stronger skeleton with resulting fewer injuries and ailments. Both sufficient available floor space and the opportunity for rabbits to jump on and off furniture, and move vertically between floor and platforms, have a positive impact on bone strength.

Rabbits with pain in their bone structure will often refuse to be handled in an effort to prevent further discomfort. They may consequently be viewed as aggressive. These rabbits are often resigned to spending the rest of their lives in a tiny cage, if not being killed as a result of bad behaviour. Unfortunately, this is only one example of how misunderstood the species is and how much they suffer as a result. These rabbits should rather be offered treatment and gradual expansion of their available area. Inactive animals will also be prone to obesity, which is a common problem among companion rabbits.

Rabbits' Behaviour and Wellbeing

What are the consequences when rabbits have to live in individual small cages that do not provide for their needs? When rabbits lack control and are unable to express their natural behaviour, they are likely to develop abnormal behaviour. A depressed rabbit will seem apathetic, a frustrated rabbit might seem

Harald, an extremely relaxed and content rabbit with free access to the apartment and veranda.
Photo courtesy of Jens Petter Salvesen, Norway

aggressive, and in both cases they are likely to use coping methods to manage their lives. Strategies for dealing with such inadequate living conditions may be reduced responsiveness or stereotypical behaviour. Stereotypies are behaviour patterns that are repetitive and have no goal or function, such as bar-biting or circling in the cage. Such abnormal stress management indicates reduced welfare, and one has to improve the rabbit's living conditions. Unfortunately, many people think it is normal for the rabbit to sit still and look depressed, because most rabbits hutched up look that way. An animal trying to cope with an inappropriate environment is thus the standard for the species' normal behaviour. The abnormal becomes normal, and it is important to raise the awareness and knowledge of rabbits to prevent this.

Rabbits are often confined in unsuitable hutches, something that deprives them from behaving naturally. Physically, subsequent frustration might lead to digestive ailments and diseases, while mental stress may result in apathy and depression.

Rabbits with an inability for being active or experiencing pleasure will inevitably spend much of their time sitting still. What else is there to do? Such apathy may well be explained by learned helplessness, a term in psychology describing that animals learn that they cannot influence the situation. Experiencing a lack of control might lead to the apparent failure to respond in future situations. The rabbit knows it is unable to escape or fight the scary hands picking it up and will just sit quietly. However, a passive and lethargic rabbit is not a cheerful rabbit. They should be investigating, be curious, active and alert, they should dig and chew, jump and run at full speed and throw themselves on their side in total relaxation.

Melissa received a second chance after 7 years of being hutched up. Photo courtesy of Hege Johansen, Norway

It is never too late

Melissa had lived alone in a hutch for 7 years when her caretakers decided it was time to euthanize her. She was left at the clinic in a horrible condition and had obviously been neglected for a long time. Her skinny body had no muscles, she was dehydrated and weak, had matted fur and could barely move. In addition she looked depressed, and we wanted her to have the chance to experience more than a hutch before she had to die.

She moved into foster care and was given the opportunity to be a free-range rabbit. However, the first days she stayed close to her litter tray and hiding spots, but eventually started to explore and move about.

After a couple of months, her body is fit and healthy, her fur is lovely and Melissa is clearly a happy and playful individual. The rabbit house has a huge outdoor enclosure, and she is digging and running like she has never done anything else. She plays in the huge outdoor enclosure and runs towards her caretaker whenever she is nearby. However, the most important improvement in life is her rabbit companion. The new couple are fooling around and this is a definite example of how it is never too late to help someone and offer them a second chance in life.

Anyone living with a free-range rabbit is aware that they have a fuller behavioural repertoire and seem happier than those kept in cages. When using activity level and observation of natural behaviour as a measure of welfare, several studies have also emphasized the fact that freedom to exercise and be with companions increases the species' welfare.

Do rabbits with access to an outdoor run show a wider variety of behaviour and level of activity than those being kept in a conventional hutch? This was investigated in a recent study,[7] which clearly showed that rabbits need space. The researchers concluded that the rabbits had far better welfare in the pen or Runaround (http://www.runaround.co.uk/) than in the hutch alone, and that the pen system was the best option to exert their natural behaviour.

To summarize, it is scientifically proven that rabbits are more active and exert a greater variety of behaviour when given the possibility. A hutch is not enough!

Rabbit Interior

In addition to offering the rabbits a larger floor area, one should also improve the available space in line with what the rabbits want and need. Enclosures, cages and other accommodation alternatives should be arranged and furnished in a way that enables the rabbits to freely express their natural behaviour.

Wide-open areas can feel threatening and scary for a prey animal, and one must offer the rabbit sufficient hideouts within the available range. Rabbits will generally feel uncomfortable in open fields when not being near a retreat. It is therefore essential that the rabbit knows where to find appropriate hiding places.

For a prey to feel safe, it is also crucial that they have a good surface to move around on. Carpets can be used to cover smooth parquet floors or similar coating so that the house rabbit has the opportunity to run off. Although the chance to meet a fox inside the house is minimal, the rabbit will always be alert to potential hazards. As their instinctive response will be to leap into the nearest cave, it is important that they have available hideouts as well as good grip on the floor.

To ensure the rabbits are able to run and move adequately, the carpets should be slip-free. It is not sufficient to have rugs that slip away when the animal tries to spin around. Heavier blankets or carpets that have a sticky underside are on the other hand a fine alternative for enthusiastic rabbits. Whether the rabbits live indoors or outdoors,

This house is designed to allow the rabbits to express their natural behaviour. Jokk and Amos can dig and run, and have access to lookouts, hideouts and even an outdoor run. Photo courtesy of Katarina Vallbo, Sweden

flooring affects the welfare of the rabbits, since poor flooring might lead to injuries from slipping, undesirable pressure on the foot, inadequate thermoregulation or exposure to moisture.

Rabbits should always have the opportunity to move up in height, either by offering custom shelves and platforms, or available tables and chairs.

Rabbits are naturally very clean, so they must always be provided with a litter tray with suitable bedding, in addition to fresh hay and water.

There are a number of stimulating and popular buildings, caves and tunnels that rabbits can play in and enjoy. The rabbit can dig in, chew on, destroy and be destructive in a cardboard castle, or just sit on the top floor, peeking out over the kingdom.

Tunnels are also recommended for rabbits. Specially designed rabbit products of untreated wood are safe to gnaw after the rabbits have lain in, dug in and run through the pipes.

Enrichment

Enrichments are defined as adaptations in the environment aimed at increasing the animals' well-being. Such environmental improvements will provide captive animals with opportunities so that they can perform natural behaviour and consequently improve their welfare. The result of different types of enrichments might be an increase in activity as well as reduction in stereotypies, such as bar-gnawing. Even fearfulness has declined in animals given enrichment. Environmental enrichment is thus a simple and effective means of improving animal welfare in any species.

Tarzan, Milo Grevling and Milla have their own room in addition to access to the rest of the house at night.
Photo courtesy of Line Leirstrand Øvrum, Norway

Frodo and Frida

Frodo and Frida. Photo courtesy of Marit Emilie Buseth, Norway

Frodo was found in a dirty cage with fly larvae and other insects. The terrified rabbit in the cage was abandoned in one of the squares in town, but was fortunately found and taken to the local rescue. After being neutered and given a thorough health examination, he quickly moved into his new home.

Frodo was petrified with fear and stayed under a chair in the beginning, but he had access to the living room, a place with his litter tray, his food and cardboard boxes, and became more outreaching every day that passed. After a while he cautiously ventured out on the floor, ran on the carpet, slept under the dining table and enjoyed sitting near the couch whenever his people sat there in the evening.

Providing rabbits with freedom of movement and a sense of control over their own bodies are key components to ensure that they are confident and happy. One has to be patient, and in the case of Frodo, he became a cuddly rabbit that followed his humans around and demanded to be groomed.

After 7 months, the foster parents have seen an impressive change in behaviour and become familiar with a new species. They decided to adopt him and give him a rabbit companion, and within a few days he was already bonded with Frida.

Frida had been observed around a train station for a couple of weeks. It was a wet and cold Norwegian winter with little grass to eat and nowhere to hide and keep dry. She was exhausted when finally rescued and slept for days. She recovered well, was neutered and adopted by Frodo.

The new couple are now sticking together like glue. They are exploring the house, grooming each other, and luckily both received a second chance.

Kelso in his Cottontail Cottage. Photo courtesy of Jannicke Torseth, Norway

Harald in his tunnel. Photo courtesy of Marit Emilie Buseth, Norway

> ### Rabbit internal accomodation
> - Ability to move between different height levels.
> - Available hideouts.
> - Good substrates.
> - Litter tray with an associated hay rack.
> - Opportunities for digging and gnawing.
> - A cage or hutch never left standing in direct sunlight.

Enrichments can roughly be divided into five groups,[8] and I will explain them based on rabbits' needs. As we will see, it is quite possible that a given enrichment belongs to several groups.

Food-based enrichments

Rabbits are herbivores that spend most of their time grazing, processing food and eating nutritionally-poor plants. To encourage this behaviour it is important to follow the guidance on diet, which primarily consists of hay, in addition to prolonging the feeding experience by hiding other food in cardboard boxes, making the rabbit hunt for vegetables hidden around the apartment or offering the daily amount of pellets in the same manner. Spending lots of time nibbling hay prevents boredom and maintains a healthy digestion. The same applies to various chewing objects.

Physical enrichments

Rabbits are, as we have seen, dependent on physical activity and exercise for maintaining their well-being and health. Good enclosure designs for rabbits incorporate both a large space and the possibility to move vertically on different platforms, and of course hideouts and chewing objects, such as tunnels of unpeeled willow or similar. Activities such as digging and gnawing should also be made possible. Digging is a behaviour that many rabbits attempt to carry out, or they indulge in displacement behaviours such as scratching carpets, etc. Providing escape-free digging areas outside, or bringing trays or boxes of earth or other substrates inside (e.g. turf, deep boxes of hay and even snow), helps fulfil their need for digging.

Sensory enrichment

Rabbits are not as visual as humans, and their main senses are smell and hearing. It is therefore stimulating and activating for the rabbit to be exposed to various fragrances and use its nose to search for food. Put some fresh herbs or vegetables here and there and let the rabbit find them.

Rabbits have good hearing, so talk to the rabbits, let them learn how to recognize your voice and tone. Rabbits can easily learn to distinguish between a cuddly voice and a strict 'no' when trying to chew on something forbidden. Offer them a radio with talk shows and classical music sometimes, and they will become accustomed to a variety of sounds.

Social enrichment

Rabbits are gregarious and incredibly social animals. Ensure your rabbit has suitable company, and

Melis with one of her chewing objects from Busy Bunny. Photo courtesy of Siv Johanne Seglem, Norway

make sure all animals are neutered, introduced properly and have a sufficiently large living area.

Cognitive enrichment

Many domesticated rabbits live in simple and monotonous environments, where they are not being challenged at all. Problem solving, exploration and play seem important to counteract boredom and apathy. Let them work for obtaining food, give them access to places they do not tend to be and see them eagerly explore the area, relocate the furniture in your or your rabbit's house and observe the busy bunny.

Free-range Rabbits

Free-range implies that the rabbits are able to move unrestricted over a relatively large area, as in an apartment or house with a secure garden. The rabbits are independent of humans and can move freely, determine when to run, play or sleep, in addition to having plenty of space to investigate and keep track of. By rabbit-proofing the home, one can let the rabbit roam free like a cat or dog would do indoors. Offer them a litter tray with an associated hay hedge, water, hideouts and places for them to feel at home and secure. The rabbits will find their favourite spots, and you might find that they prefer to be in various places at different times of day.

Rabbits like to be able to climb to get an overview. Normal furniture like sofas, chairs, tables or even a piano can serve as lookouts.

Many owners allow the rabbits to be free-ranging throughout the house, while others have restrictions and let them be in different enclosures at night or when people are out of the house. Some rabbits are also referred to their own rabbit-secure rooms when they are not under supervision.

Harald hunting for herbs. Photo courtesy of Marit Emilie Buseth, Norway

Fencing attached to the wall. Photo courtesy of Marit Emilie Buseth, Norway

Enclosures

There are many fences that can be used for making indoor enclosures. Puppy pens, child playpen and compost heap fencing are examples of what can be used to expand the rabbit's living area and prevent access to the rest of the house or just some exposed furniture.

Be aware that such fence materials are not intended to enclose small animals. The holes may be large enough for young or small-breed rabbits to either crawl through or become stuck and injured. If one has tiny rabbits, it is therefore recommended to secure the lower part of the fencing with a fine metal mesh.

Such a fence system is easy to set up. With the help of strips one can connect a desired number of elements, put up the series with fences and then furnish the new pen.

If the fences end in a circle, it is easy hooking them together. If the end pieces should be attached at different locations, this can easily be solved by screwing a hang tab into the wall with a subsequent connection with strips.

'NIC grids' and 'storage cubes' are types of mesh storage systems that are also widely used as rabbit housing. They are smaller in size and therefore more flexible in terms of building different sections and levels, and are available commercially from storage solution vendors, as well as some pet shops.

Further improvement can be achieved by building in height. Giving access to higher levels allows both more space and greater environmental complexity and choice. Rabbits may enjoy lying under shelves, as well as on top of them. However, it is important to ensure that the shelves are properly fixed, as insufficiently wide or unstable shelves or ramps may prove hazardous to the rabbit.

Allowing choice permits the rabbit to select an area in which it feels comfortable, which often is difficult for humans to predict, so if they can vary between floors, they suddenly have a more suitable living area.

Dolly. Photo courtesy of Marit Emilie Buseth

> Compost heap fencing or similar pens are not suitable for permanent housing outdoors. However, as a run under supervision, they work fine. See Chapter 11 for information on secure outdoor housing.

An indoor enclosure as a base. Photo courtesy of Katarina Vallbo, Sweden

Our most used compost fences are 68 and 88 cm in width and height, respectively. Most rabbits remain within the allocated space, while a few seem to jump out of the pen. One must be aware that certain rabbits will be able to jump over such heights. If you have an especially energetic and keen animal, it may therefore be necessary to roof the compost wall system. This can be done by connecting several fences and attaching them in the same way that the rest of the enclosure was made.

The size of the fences makes them suitable to put in the doorframe if you want to prevent access into various rooms.

Cages and Space

Cages designed specifically for rabbits are generally unsuitable to be anything other than a base for litter trays and hay, and an area to retreat into at will. Most cages sold in pet stores and other retail outlets for farmers have no opportunities for enrichment or room for exercise and thus are not suitable as permanent residency.

All rabbits confined to such a cage must have an attached exercise run. They should have permanent access to a room or an enclosure, or at least have access to it for a minimum of 4 hours a day, preferably at times when the rabbit is naturally active.

The previously mentioned fencing can easily be set up around the hutch or area, and thus act as an extended living space.

The Rabbit Welfare Association & Fund (RWAF) have a campaign whose goal is to increase the size of the cages sold in stores. The project, called Bigger is Better for Busy Bunnies, will demand that the minimum requirement for a cage or hutch must be at least $1.8 \times 0.9 \times 0.9$ m. The stores will also have to sell much larger cages and enclosures. When they also need to have room for a litter tray and a house, cages need to be larger than those generally sold in pet stores.

As a prey species, rabbits are dependent on being able to stand on their hind legs and prick up their ears without touching the ceiling. They orient themselves and communicate by means of their ears, and if they do not having this opportunity, their welfare is reduced.

Scandinavian design: the Hol table from IKEA is perfect for an indoor rabbit. The shelter may be used as a place for the litter tray and hay rack, or it might be a popular cave, as in the picture. Two entrances are recommended. Photo courtesy of Marit Emilie Buseth, Norway

Another view of the Hol table from IKEA with a litter tray and two hungry rabbits inside, Goliat and Lexie. Photo courtesy of Ingeborg Tjore, Norway

A much used guideline for space requirements is to provide enough for the rabbit to take three hops in a straight line. This, however, has been both overlooked and interpreted to the disadvantage of the rabbit, and many seem to underestimate the rabbit's abilities and needs. Based on these guidelines, an average rabbit will in practice need a cage which is 2.0 m long. These requirements derive from laboratory animal welfare guidelines and should be considered an absolute minimum, as they do not permit normal exercise. Rabbits should be able to run rather than simply jump, and one should keep in mind that the species, despite their relatively

Many rabbits spend their entire life in small cages. However, Alvin is one of those fortunate few who only use it as a base for food, shelter and his litter tray. Sixteen-year-old Anne got help from her parents to hide electrical cords and wires, so that Alvin could safely enjoy free access to all of her bedroom.

Alvin is neutered and clean. He either uses the litter tray, which is located inside the base, or the one that is placed outside, both with related hay hedges. To avoid fur or accidents in bed, compost fences protect this when Alvin is home alone. The couch is only interesting when Anne is staying there, and Alvin clearly likes to seek her out for company and mutual grooming.

It has been easy and trouble free to allow Alvin to be free-range. He enjoys the freedom of movement, and it is also more fun and easier for Anne to have a rabbit in this way. She will not have to change and wash filthy cages. If you have free-range rabbits they will use a litter tray, which is far more convenient and easy to change when needed.

The dog cage serving as a base gives the rabbits possibilities of movement and choice of platforms, standing upright in full length, and the possibility to live with a companion rabbit. Photo courtesy of Bunspace.com

small size, needs plenty of space to move freely. They do not wander, like guinea pigs or cats, but have a more space-demanding way to move.

The largest dog cages are practical as a base if wanting an alternative to more traditional cages. They are also easier to build in height and can be extended by connecting multiple cages. One can easily insert platforms so the rabbits can move between different levels.

A module-based storage system, like the above-mentioned enclosures, makes it is easy to build custom-designed cages. One can build in height and width and design a better-adjusted rabbit residence with different floors and compartments. It is easy to set up a three-storey variant, offering different facilities at each level, and ensure that the rabbits have the ability both to initiate and draw themselves away from social contact.

Regardless of the type of caging, rabbits should have permanent access to an exercise yard or free movement outside the cage. Although rabbits can move between different floors by jumping up on platforms, they still need to find an outlet for other movement. Besides their need for exercise, rabbits must be able to experience and express joy and happiness, as when running like crazy when having a raptus (running and turning in a zig zag fashion) or binky.

A customized rabbit room in the house. Photo courtesy of Aksel Hunstad, Norway

Air Quality

Poor air quality is less of a concern with rabbits cohabiting with humans than those outside in small confined areas or group-housed. Ammonia and dust levels may be higher at rabbit level than human head height, but otherwise human inhabitants will not notice poor air quality issues. Smoking appears to be a risk factor in the development of cardio-respiratory disorders in rabbits, and should be avoided in the house.

> ### Problems with rabbits kept in breeding systems and sheds
>
> Some rabbits, typically in breeding or showing establishments, are housed in stacked or other multiple hutches outdoors or in sheds. The main problems with this are lack of space and social housing. In addition it is often difficult to maintain adequate biosecurity and to achieve a balance between good hygiene, temperature control, humidity and ventilation.
>
> Typically, such rabbits are permanently or near permanently housed in hutches by themselves, or, in the case of does, with their litters until weaning. They may be given access to outdoor run areas on a rota basis, or may have their own run space, which they may be placed in during the day. It is rare that they have permanently attached outdoor runs.

> Sara was 11 years old and thrilled because she was finally allowed to get a rabbit. She was very excited when they drove off to buy a rabbit, and they had been assured that it was a reputable and one of the most experienced breeders they were going to meet. Sara shook hands with a friendly woman before entering a huge hall, but the sign and smell shocked her. Over a hundred rabbits lived there, many looked anxious, while some seemed aggressive towards rabbits in the nearby cage. Most rabbits lived alone in their small and naked cages, while all mothers lived with their children, often with several litters in the same hutch. Sara had a bad feeling but was told that this was perfectly normal.
>
> *Continued*

Continued.

Sara was specifically aware of a small female rabbit that looked very scared and nervous, paid and brought her home. She was very beautiful but incredibly frightened, and Sara regretted the moment they arrived home. The rabbit seemed angry and Sara didn't know what to do, besides keeping her in the hutch. The breeder had after all described this as normal rabbit husbandry, and Molly, as the rabbit was called, sat consequently in the corner for the next 6 months.

Fortunately Sara realized that this was a sombre existence for the rabbit, and when she got in contact with me she wanted Molly to feel better. After discussion, the rabbit was offered a bigger cage with an attached enclosure. Sara spent lots of time there and it was easier to socialize the rabbit. Although Molly remained difficult and scared she improved rapidly.

Nevertheless, it was not until she moved into a larger dollhouse the real improvements were visible. She turned into a totally different rabbit, and Sara tells me today that Molly is a more secure, straightforward and a curious and active rabbit. Sara can cuddle her and handle the previously nervous and aggressive rabbit. After turning 13 years old Sara has now been allowed to move the rabbit indoors to be a free-range rabbit in her bedroom. Molly seems to enjoy and benefit from the more social environment and Sara can develop an even closer relationship to her dear friend. Currently Molly has even accepted a companion rabbit.

Notes

[1] Berthelsen, H. and Hansen, L.T. (2000) The effect of environmental enrichment on the behavior of caged rabbits (*Oryctolagus cuniculus*). *Applied Animal Behaviour Science* 68, 163–178.

[2] Held, S.D.E., Turner, R.J. and Wooton, R.J. (1995) Choices of laboratory rabbits for individual or group- housing. *Applied Animal Behaviour Science* 46(1), 81–91.

[3] The Animal Welfare Act (2006) Available at: http://archive.defra.gov.uk/foodfarm/farmanimal/welfare/act/documents/aw-act-2006-memo-101220.pdf (accessed 25 February 2013).

[4] Schepers, F., Koene, P. and Beerda, B. (2009) Welfare assessment in pet rabbits. *Animal Welfare* 18, 477–485.

[5] PDSA (2012) PDSA Animal Wellbeing Report. Available at: https://www.pdsa.org.uk/pet-health-advice/pdsa-animal-wellbeing-report (accessed 9 November 2012).

[6] Drescher, B. and Loeffler, K. (1991) Einfluβ unterschiedlicher Haltungsverfahren und Bewegunsmöglichkeiten auf die Kompakta der Röhrenknochen von Versuchs- und Fleischkaninchen. *Tierärztliche Umschau* 46, 736–741. Cited in Tettamanti, M. and Veeraraghavan, P. The impact of facts from the rehabilitation of laboratory rabbits on reliability and evaluation of experimental data. Available at: http://www.icare-worldwide.org/images/poster_4_rabbits.pdf (accessed 2 November 2012).

[7] Redrobe, S. (2011) Is a hutch enough? A comparision between hutch only, hutch & pen and hutch & runaround systems. *Proceedings from The 2011 RWF Conference*, Solihull, 29 October 2011.

[8] Melfi, V. (2010) Enrichment. In: Mills, D.S., Marchant-Forde, J.N., McGreevy, P.D., Morton, D.B., Nicol, C.J., Phillips, C.J.C., Sandoe, P. and Swaisgood, R.R. (eds) *The Encyclopedia of Applied Animal Behaviour and Welfare*. CAB International, Wallingford, UK, pp. 221–223.

Petter on his piano. Photo courtesy of Marit Emilie Buseth, Norway

10
House Rabbits and Rabbit-proofing of the Home

Wall-E and Eva are free-range house rabbits. Photo courtesy of Marit Emilie Buseth, Norway

Allowing the rabbits to take advantage of the same benefits that cats and dogs have for years is a relatively new phenomenon and unknown to most people. The more privileged relatives of wolves and tigers have historically achieved status as man's 'best friend' and distinctive room-mates, while rabbits have been favoured with some bread crumbs in the yard.

Traditional farming of rabbits involves housing in hutches, slaughter when their fur and flesh are of the best quality, and little interest in and knowledge about their welfare. Rabbits have, since people began to make use of them, had a long story as cheap, easy and effective livestock. This has probably affected people's attitudes, and even today many still do not see the species as a potential companion animal.

One of house rabbits' great advantages is that human and rabbit know each other's habits and way of being. When you live close to each other, it is easier to detect deviations from normal behaviour in the animal, which may indicate discomfort and disease. It is easier to identify symptoms, and decisive treatment can be started quickly. House rabbits thus have an advantage compared to their relatives in the garden, which in many cases are only visited once a day. Rabbits are subtle beings, and there are usually tiny changes that lead one to recognize something is wrong. Vague signals can be difficult to detect if one only sees the rabbit when serving a bowl of food. (See list of symptoms on p. 93.)

Daily life is characterized by routines. This is also ideal for rabbits who appreciate predictability and stability. It may seem as if they have clocks, as they know when it is time to be served a salad for breakfast and when pellets should be offered in the evening. Many rabbits will clearly tell when it is time to get up in the morning, often by scrabbling at the door or jumping up on the bed to make sure they awaken their servants.

In periods where the temperature and climate conditions are suitable, rabbits can obviously live comfortably outdoors. However, their wellbeing is dependent on their ability of freedom of movement, shelter and companionship of other rabbits. When one does not have the ability to provide a satisfactory living environment outdoors, I recommend a cohabitation inside the house, where the family pet has better opportunities for regular exercise and social interaction, rather than a rabbit sitting alone in a hutch and being something that no one seems to remember when the weather is bad or there is something good on TV.

Illustration courtesy of BunnyHugga.com

A babygate is set up to prevent access and passage of the stairs. Bella inspects the photographer. Photo courtesy of Marit Emilie Buseth, Norway

Bella and Lillos have a custom-made room with their litter tray, water and hay under the stairs.
Photo courtesy of Marit Emilie Buseth, Norway

It normally requires less work to have a house rabbit than to keep one outside in a hutch. In addition, it is naturally more joyful. The rabbits will adapt to the household, and aligned with dogs and cats they become part of family life. One will also enjoy mutual company, even when the rabbit is resting on the floor while the human is doing its homework or making dinner.

Rabbit-proofing Your Home

Most accidents occur within the home. For the same reason that we make the house baby-safe, or secure it against a destructive pup, or even set up a scratching post for a playful kitten, one must also ensure that the rabbit lives in a safe and suitable home.

The electricity challenge

Most people living with house rabbits are greeted with some surprise and must at times answer some standardized questions. The majority of concerns are usually of an electrical character: 'Do they not eat wires?'

Yes, most rabbits are attracted to electrical cords, and one must therefore ensure that they are out of reach. Never trust your rabbits to leave the cables alone, even if they behave properly when you are around. There are fortunately many ways to prevent electrocution and other expensive and dangerous situations.

Continued

House Rabbits and Rabbit-proofing of the Home

Continued.

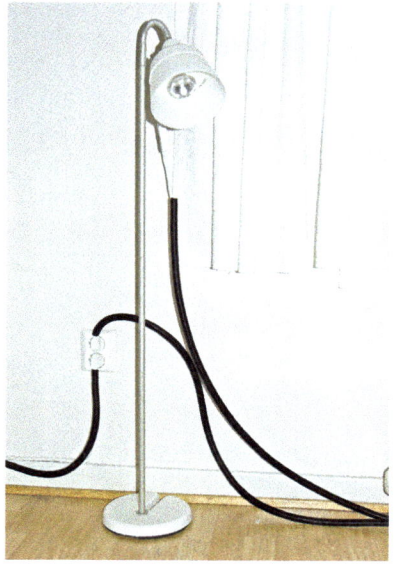

Flexible protection. Photo courtesy of Marit Emilie Buseth, Norway

Flexible spiral cables and wire channels

Wires can be protected with flexible cable wraps or covered in channels attached to the wall.

Flexible protection

Plastic tubing or spiral cables can be obtained in different diameters from most hardware stores, and because the wires will be manageable, they are well suited to protect the vacuum cleaner, which is widely used in any rabbit home.

These flexible cable wraps are perfect for lamps or electrical equipment that is likely to be moved around.

Be aware of exposed areas such as the transition to the AC power outlet.

Permanently mounted protection

Wire channels of plastic, or PVC for determined rabbits, can be attached on to the wall or mouldings.

An electrician can make direct input to the AC power outlet so you avoid the above-mentioned problem.

Permanently mounted protection. Photo courtesy of Marit Emilie Buseth, Norway

Mouldings with room to hide wires are available in plastic or PVC.

Location of the power socket

Sockets can be mounted higher up on the wall so that the cord in use becomes unavailable for the rabbits.

By placing the sockets higher on the wall, they are also easier to hide with the help of furniture, such as shelves and sofas.

Protective interior

Media benches with room for wire storage can often hide large amounts of cable. If you have a normal cupboard, one can also build an improvised media bench by making a hole against the wall and on the top. Then you can safely put the wires in the electrical outlet, pull the cord out of the hole and put it in the TV. The cable will then be concealed by the furniture.

Change of mind

With a rabbit in the house, one has to think differently. One can no longer leave a mobile charger on the coffee table or leave an unattended vacuum cleaner with wires hanging. Telephone cabling should be covered as well. Be vigilant during activities that require a power supply, such as ironing, vacuuming or charging a PC or mobile phone.

Keep the rabbit in another rabbit-proofed area whilst vacuuming if necessary.

Woodwork

Rabbits are rabbits, and they will always have the need to chew and dig. Some are more eager than others; those less than a year old can be particularly challenging. Just as a puppy will chew on furniture, eat shoes and urinate on the floor, a young rabbit will also be an investigative fellow who will learn along the way and adapt to family life, and for the same reason that a cat will have a litter tray and toys to scratch on, a rabbit should also be provided a suitable living environment. To protect wooden furniture, cover the legs with cardboard or plastic tubing.

Distractions

To ensure that the rabbit's natural needs do not cause major damage to table legs, mouldings and cabinets, make sure that other chewing materials are present. Cardboard boxes, twigs, branches, chewing products and other activities must always be available.

Mouldings

If your rabbit shows great interest in woodwork mouldings, you can replace them with the above-mentioned plastic or PVC mouldings, or cover them up with channels, tubes or other material, such as Plexiglas.

Many rabbits also seem to detest mouldings or fences covered with shiny shoe polish.

Wallpaper

Wallpaper can be surprisingly fun to tear off and eat. If rabbits are eager to put their teeth in the wallpaper, it is possible to prevent this by covering areas with transparent material, such as Plexiglas or plastic. Fasten a plate on the wall at a suitable height.

Corners

If your rabbit has found himself a pleasant corner to gnaw on, it is possible to secure it against further destruction by turning a corner protector.

Blankets

In the same manner as rabbits need to chew, they also need to dig. Some will dig away at selected carpets, especially in the corners. If one has wall-to-wall carpeting, one can therefore staple in small pieces of carpet samples in selected corners, so the rabbit can scramble and destroy without damaging your carpet and spirit. Ceramic tiles or other protective material are also useful to have in the corners.

Be aware and make sure that rabbits do not eat the carpet, or carpet underlay, as indigestible material can lead to blockage of the digestive system.

Plants

Houseplants should be kept out of reach. They will be tasted or eaten and many of them are toxic to rabbits.

If your rabbit has access to an exercise yard in the garden, one must also ensure that there are no harmful flowers or bushes.

Other toxins

Other potential toxins include prescription medications and heavy metals. Rabbits are perfectly capable of chewing through foil packaging or plastic bottles to get at human or animal medication, but seem to rarely do so. Should this happen, contact your veterinary practice at once, with full details of the drug eaten, including the amount, the time, and the exact name and size of the tablets. Keeping the product insert information is wise.

Rabbits may eat heavy metals, including lead and zinc. Lead is not widely used nowadays, but may be found in old paintwork, old lead toys, old lead piping, leaded window fittings, angling weights, and bullets and pellets. Zinc may be found in some wire mesh (typically the light grey wire with droplets of metal at the corners of the mesh gaps), and this must not be used. If heavy-metal toxicity is suspected, fluid therapy and a specific antidote are used to treat it.

Delimited areas

To limit the rabbits' access to parts of a room or the lowermost books in the bookshelf, one can easily make use of different fences or other suitable plates. There are many fences that can be used for making indoor enclosures. Puppy pens, a child's playpen and compost-heap fencing are examples of what can be used to expand the rabbits' living area and prevent access to the rest of the house or just to some exposed furniture.

Be aware that such fence materials are not intended to enclose small animals. The holes may be large enough for young or small-breed rabbits to either crawl through or become stuck and injured. If one has tiny rabbits, or tempting wires are within reach, it is therefore recommended to
Continued

Continued.

The bungalow from Oxbow is nice to both sleep in and eat at, as Petter is illustrating. Photo courtesy of Marit Emilie Buseth, Norway

secure the lower part of the fencing with a fine metal mesh or other concealing plates.

Walking through closed doors

Be aware of fast rabbits sneaking in through the door without you even noticing it. I have even closed rabbits inside the bedroom, out on the veranda or inside a closet, so one should always make sure that all individuals are where they are supposed to be when you leave the house or go to bed at night.

Another clever rabbit I know walked through closed doors. Motivated by biscuits and a desire to explore, Pixel found her way into the kitchen

Continued

Continued.

Cottontail Cottage is a popular hideout. Photo courtesy of Marit Emilie Buseth, Norway

Corner protection. Photo courtesy of Marit Emilie Buseth, Norway

Delimited area. Photo courtesy of Marit Emilie Buseth, Norway

cupboard, whereupon she ate a packet of biscuits and some spaghetti. Due to the ensuing bad stomach, she was fortunate to be living with two veterinarians.

Few households are equipped with qualified health professionals, so it is important to limit access to undesirable foods.

Food and beverages

Everything within reach can be tasted, destroyed or eaten. Bowls with chocolate, cakes and sandwiches in particular must therefore be monitored continuously or be beyond possible attacks. A cup of hot tea can inflict great damage on

Continued

Continued.

Bowie Bunstructor. Photo courtesy of Karen Kowalaski, Pennsylvania, USA

Be aware of the rabbit! Photo courtesy of Marit Emilie Buseth, Norway

both textiles and snouts. Eager paws must therefore be kept at a safe distance.

If one eats while sitting on the couch, one must be aware. Some rabbits will sneak up on and steal the human food, take a mouthful and run away.

The toilet seat

Please note that some rabbits can jump in the toilet, which will be disastrous if it happens when you are asleep or not at home. Always keep the toilet seat closed.

Shoes and chargers

Until you know the rabbit's habits, you should ensure that shoes, bags and sacks are stored in a suitable place. To prevent nibbling on the said products, any visitors should also be informed of these security measures. They should also be told that mobile chargers cannot be left on the floor, the coffee cup cannot be left unguarded on the coffee table, and by all means do not step on the rabbit!

Mandel illustrates how important it is to have both hideouts and items to gnaw on. This bench serves as both. Photo courtesy of Marit Emilie Buseth, Norway

Mia and Milli are relaxing in the tent.
Photo courtesy of Marit Emilie Buseth, Norway

11

Life Outdoors

Wild rabbits obviously live outdoors. However, traditional housing outdoors, where rabbits are hutched up alone, has nothing to do with the species' natural way of life, as this outdated husbandry prevents innate needs such as social interaction, running and activity in general. Fortunately it is easy to offer them a better life with some knowledge and commitment, and this chapter will address necessary adaptations and provide practical solutions.

Not everyone has a garden or outdoor area for rabbits to use. Different countries also have various challenges, and in some climates it is not recommended to have rabbits outdoors for certain times of the year, because of intense heat or cold. In other countries, there may be challenges related to parasites and predators. Many people can more easily provide sufficient living areas inside, and it is therefore important to make people aware of how to prepare for having rabbits inside, for at least part of the year (see Chapter 10).

Mia enjoying her huge run. Photo courtesy of Marit Emilie Buseth, Norway

Adaptation in the Wild, and Optimal Temperatures for Rabbits

It is important to protect domesticated rabbits against extreme temperatures, both from the heat in summer and cold in winter. Wild rabbits live in underground tunnels, where they are not exposed to large variations in temperature. In the summer, the burrows remain relatively cool, while they are sufficiently warm and dry during winter. This ensures that rabbits are adaptable to the seasons.

However, the problem arises when we take the rabbits out of their natural habitat and keep them in cages. The hutch is above ground level and ensures that the rabbit is exposed to all kinds of weather and temperature, from which in the wild they would have sought refuge. Rabbits do not tolerate humidity, draughts and damp conditions, and they must always have access to sheltered caves or houses.

It is therefore our responsibility to ensure that rabbits are offered optimal conditions. In Norway for example, there are challenges in terms of keeping rabbits outside all year, due to harsh and freezing winters, while Australia will experience problems concerned with the heat in summer. Even in the UK, temperatures may exceed the comfortable limits for rabbits. The optimal temperature for keeping rabbits is 15–20°C.[1]

Rabbits in the Winter

Despite the ideal temperature conditions, rabbits tolerate degrees below this range relatively well. However, it is essential that some precautions are taken to ensure the outdoor rabbit's health and welfare.

Regardless of where rabbits live, they must be kept free of damp and draughts, and always have

Gråtass always runs towards humans, while his companion rabbit is more restrained. Rabbits are individuals and some are more interested in social interaction with humans than others. Photo courtesy of Marit Emilie Buseth, Norway

access to a secure area to retreat into. A cold and draughty environment could lead to serious respiratory infections, in addition to suffering in the rabbit's joints, muscles and skeleton. To ensure that the rabbit does not develop osteoporosis and consequently fractures and pain, rabbits should have ad-lib access to outdoor runs, so that they can maintain a good body condition.

Rabbits without an attached run are otherwise dependent on being allowed into them by owners. This is likely then to only occur during daytime, in good weather and when the caretakers have no other plans. Knowing that rabbits are naturally active during twilight makes this a challenge. Their wild predecessors are playing and grazing outside in the evening through to early morning, and sleep in their burrows most of the day. A brief run after school hours is thus not ideal for the nocturnal animal.

Fresh liquid water must be available at all times. Water should obviously never be prevented from freezing by adding chemicals such as antifreeze or salt, so frequent replenishment is necessary at temperatures lower than freezing, and water bottles may be insulated with home-made or commercially available insulating jackets. Ensure that the nipples of drinking bottles have not iced up. Please note that the rabbit's ear might fall in a water bowl and become wet, which can result in frostbite.

An insulated house can be filled with straw, but at temperatures lower than freezing, supplementary heating might be helpful, such as microwavable heat packs placed in the sleeping area, so that the rabbit can be close to it at will.

Rabbits may spend more time inside the house than usual, so remember to replace the bedding and clean more frequently than usual, so that the living area is kept dry and clean.

In countries with severe cold conditions one should also have heating in the rabbit house. Electric heaters should be dustproof and well insulated from any risk of water entry or of wires being chewed. Other heaters should be safe, protected against risks of causing fires, falling over or emitting dangerous fumes. They should maintain a steady and gentle temperature of 5–7°C. If the temperature is set higher than this, it will be uncomfortable for the rabbit to stay inside due to the variation in temperatures between the inside and the outdoor run.

It is easy to mount a cat flap so that the rabbits can go in and out whenever they want. A windbreak outside is also recommended, so that the house is protected from the wind and driving rain, and the door is not in danger of being pushed shut by any snowfall. This should not be so close or tight as to prevent good ventilation. If a rabbit house is placed directly on the ground, insulation is necessary underneath to prevent heat loss, and waterproofing

A perfect house and run for rabbits who live outside all year round. The trio can be active, and they have a little door into a huge and insulated house. Photo courtesy of Kristine Røn Gisholt, Norway

is vital to prevent damp entering the house. The floor should slope very slightly towards the front of the house to avoid water, rain that has entered or urine pooling at the back. However, in countries with severely cold conditions, the house should additionally be raised up off the ground, which assists with reducing heat loss, keeps the hutch out of pooled water and offers some protection against predators. Access to the house should be via a wide, stable, non-slip slope that is not too steep, to allow even elderly and mobility impaired rabbits easy safe access.

Rabbits do not tolerate rapid changes in temperature and will struggle to adapt to differing outdoor and indoor climate during winter. Do not move the rabbit back and forth excessively, as the rabbit will find the changes of environment stressful. Rabbits require exercise all year round, so one should make sure to have a sufficient run.

Alternatively, one can keep the rabbit in a cooler area of the house for a while, such as a chilly basement. If one wants to move an outdoor rabbit inside permanently, this is no problem, as long as one ensures the rabbit has a cooler room for the transition period.

For the same reasons one cannot put a house rabbit outside when it is freezing. When the temperature is above zero they might come outside to have a run on the veranda or in the garden, as long as they keep moving and can run inside by choice. However, indoor rabbits should never move outside before summer, not until you can tolerate staying outside in a thin jumper.

Being a formerly inexpensive livestock animal, rabbits have a pretty low status in general. Typical husbandry is unfortunately still affected by a lack of knowledge and little willingness to contribute to the species' welfare. In Norway some people believe that rabbits can tolerate sitting in small wire cages at −30°C, because this has been the reality for many years. In the UK, the concept of a rabbit living in a hutch at the bottom of the garden is still widely held, as this was the traditional method of keeping them for food production. However, tradition is not a good standard on knowledge, and survival is not necessarily an indication on welfare.

Rabbits in the Summer

Rabbits do not have sweat glands in the skin, other than a few on the lips, which means that they cannot sweat. They also cannot pant, as they breathe through the nostrils, and are therefore at risk for heatstroke at high temperatures and in direct sunlight.

The species does not tolerate high temperatures and should never be confined in hutches exposed to sunlight without the possibility to move about to find cooler areas.

In countries and areas where the temperature gets extremely high, rabbits should move indoors during summer. A house in such a climate will often be ventilated and consequently provide a cooler and more pleasant environment than the garden.

Symptoms of heatstroke can be rapid respiration or the rabbit gasping for breath, loss of appetite, bluish or grey lips, and a generally exhausted and hot rabbit. If one suspects that a rabbit is hit by heatstoke, the rabbit must be wrapped in a towel lightly moistened with cold water. The towel should not be soaked but only be comfortably humid and chilly. A rabbit affected by heatstoke should never be put

Rabbit house with a huge run. Photo courtesy of Laila Vatland, Norway

Martin running into the house via the windbreak. Photo courtesy of Aksel Hunstad, Norway

Charlotte and Stein have shelter in a house above ground. Photo courtesy of Aksel Hunstad, Norway

A rabbit house above ground. Photo courtesy of Twigs Way, UK

Lars stretching in the garden. Photo courtesy of Kristine Røn Gisholt, Norway

in water as this might provoke shock. Being the rabbit's thermostats, ears should not be cooled down, as this might interfere with the regulation of body temperature.

The rabbit's hindquarters should be examined three to four times a day in warm periods, to avoid development of flystrike. Signs of wet faeces or soaking of the skin with urine must be removed immediately, and the rabbits must be offered a diet rich in fibre to prevent the onset of digestive problems. The droppings should always be round and nearly porous, while the caecotrophs should be eaten directly from the anus or immediately after if the rabbit leaves it on the floor in advance of eating it.

Humid and dirty substrates will pave the way for a fly larva attack and must be avoided. Clean the litter tray daily during summer (for more information on flystrike, see pp. 101–102).

Life Outdoors

Checklist for outdoor rabbits

- Outdoor rabbits should always live with a rabbit companion.
- Rabbits should always have access to an insulated and dry house.
- Rabbits must be protected from weather and extreme temperatures.
- Rabbits must have access to a run.
- Rabbits must be protected from predators.
- Rabbits must always have fresh hay and water.
- Rabbits should never stay on wire floors.

Winter specials

- Rabbits should not be taken inside and outside excessively.
- Never move a house rabbit outside during winter.
- Fresh liquid water must be available at all times.
- Rabbits sneed an insulated house and access to a run.

Summer specials

- Rabbits must be sheltered from direct sunlight to prevent heatstroke.
- Clean the litter tray daily during summer.
- The rabbits' hindquarters should be examined daily to prevent flystrike.

Pika cooling down by the air conditioning. Photo courtesy of Ken Kitamura, Toronto, Canada

Rabbits and Grass

Rabbits should always have unlimited amounts of hay, even when they live outside and have access to grass. However, fresh grass has the advantage of a higher water content, helping to dilute the urine and reduce the risk of urinary tract problems, and is more palatable to rabbits than all but the best hays. Grass is also slightly more abrasive to the teeth than hay, helping to wear them down. And finally, it is free!

Since rabbits do not tolerate sudden changes in diet, please note that they must be accustomed to fresh grass over a period of time. Many rabbits have become ill after eating quantities of lawn grasses when they are not used to it, so pay attention for the first few days outside, especially if this is in the spring, when the new growth of grass is very lush and protein rich.

Never give rabbits grass that has been cut by the lawn mower, and do not let them out on a newly cut lawn without removing the clippings. Clipped grass will quickly become fermented and cause digestive problems if eaten.

For the same reason one should not pick more grass and greens than the rabbits will eat immediately.

Lucy. Photo courtesy of Yvonne Wollertsen, Norway

Freezer blocks or bottles of water can be stored in the freezer and presented for rabbits on warm days. Wrap the frozen items or water bottles in towels and put them near a hot rabbit.

Rabbit Housing Outdoors

Regardless of climate and temperature, one must ensure that the rabbits have suitable and safe housing. In addition to a secure run, the animals must have access to a safe area for sleeping and retiring to for rest and shelter.

Wild rabbits live in networks of burrows, called warrens, where they live together in colonies. The tunnels can be very long and sophisticated, with a number of entrances, which explains why domesticated rabbits also seem to like houses with more than one exit.

A typically small and commercially available hutch is not enough, but it is fortunately easy to provide far more suitable living accommodation for the rabbits. A larger than standard hutch may work well as a sheltered base when combined with a permanently attached run; however, home-made and customized houses or other set-ups adapted for rabbits are preferable. A child's play house or a shed in the garden may also be a good base for rabbit housing. Install a cat flap and let the rabbit go in and out at will.

Heidi and her husband got Pelle for their 7-year-old son a couple of years ago. Due to misleading information from the pet shop they ended up feeding him museli mix and breadcrumbs. They thought it was fine to keep him in a tiny hutch, even thinking the two floor cage was luxurious, and let him run in the garden only occasionally. Fortunately, they found the Facebook group for *Den Store Kaninboka* (the Norwegian book that *Rabbit Behaviour, Health and Care* is based upon). They read *Den Store Kaninboka* and understood that Pelle had to get out of the hutch. Pelle was neutered and acquired a new friend, Lykke. She was adopted from the local rescue and they became best friends. The rabbits now enjoy this newly built, isolated house with access to a huge run. Both rabbits and humans in the family are much happier today.

Continued

Life Outdoors

Continued.

Knowledge leads to welfare. Photos courtesy of Heidi Nilsen, Norway

Rabbit housing should be held off the ground to prevent cold floors, and a ramp down will give access to the run. Large shelters will allow rabbits to get out of sight of predators and consequently feel more secure. In addition, large sheds are less susceptible to heat build-up during summer, and will ensure the rabbit has more space in cold and wet weather.

Mesh-fronted hutches offer no protection from extreme weather, predators or other curious visitors. Although the rabbits are safely kept in a cage, they will still be terrified if a fox or cat is hanging on the mesh, which may result in fatal stress reactions.

Parasols and other shades may provide protection against wind, excessive sun and rain. Be aware that plastic sheeting on the front of a hutch results in lack of ventilation and poor air quality.

Wire-mesh flooring

Rabbits should never have wire-mesh flooring in their cage. The foot anatomy makes the rabbit vulnerable to damage if living on wire floors, and they can suffer from sore hocks and broken claws. It also prevents normal movement. Such substrates will obviously not insulate, and rather cause the rabbits to be exposed to cold, moisture and draughts. In addition, the rabbits will have problems eating their caecotrophs. Normal behaviours, such as digging, are also difficult to perform on mesh flooring. Studies have also shown that animals prefer solid flooring to grid floors.[2]

What kind of wire mesh?

There are many retail outlets that sell various ready-made set-ups with both sufficient houses and attached runs. The Rabbit Welfare Association & Fund recommends a 6 ft × 2 ft × 2 ft house with an 8 ft run as an absolute minimum, but it is fortunately easy to find and build a larger and even more rabbit-friendly solution.

When building a basic rabbit set-up, it is important to use appropriate materials. The wire mesh used must be fox-proofed, and bird netting or other chicken wires are not sufficient as rabbits can gnaw their way out or predators or wild rabbits can easily enter the run. Galvanised mesh is harder and not as easy to work with but will ensure rabbits against intruders if fastened well enough.

The base of the door should be a few inches above the ground to prevent freezing and problems with snow in winter. Photo courtesy of Marit Emilie Buseth, Norway

Hugo relaxing. The run is Hugo-proofed with mesh on the ground near the border. The run is huge, so he is digging other places instead. Photo courtesy of Hege Johansen, Norway

The run

When building a rabbit set-up, one needs to remember that the European rabbit (*Oryctolagus cuniculis*) is known for digging underground burrows. It is therefore necessary to protect the run from rabbits digging out and from predators or wild rabbits sneaking in.

One can insert mesh or fences into the ground vertically, so that the rabbit meets a wall if trying to dig itself out. Alternatively one can cover the whole run with wire or fences, but if lying horizontally it needs to be sunk under the grass so as not to hurt the rabbit's feet.

Be aware of corners, as they seem to be exciting to explore. Holes should be closed and the run should be investigated every day.

Ceilings to keep predators and birds of prey out should cover the run. The roof may consist of galvanised mesh that can be attached in the corners with wire or strip. The roof should be further reinforced by studs, since snow can be heavy and destroy constructions.

If the run is missing a rooftop, the fences must be a minimum of 2 m high to prevent access for foxes and cats. Fences being used for dog yards may be suitable, especially with an angled overhang.

Check that doors are strong and add hasps and locks on the openings for extra security. The doors should be fastened such that the base of the door is a few inches above the ground to prevent them freezing to the ground in winter, or becoming difficult to open due to snow or hay or other items in the pen getting in the way.

Exercise run

A convenient and simple solution for an exercise run is to use different fencing systems to set up a suitable pen in the garden. Such mesh runs should be firmly affixed by the use of tent pegs. However, the fences are too low and unstable to serve as a rabbit-proofed enclosure. It is not considered safe and should only be used under supervision. In addition, such provisional enclosures have no roof or securing into the ground, potentially allowing the rabbits to get out and predators to enter.

A hiding spot must always be available in the exercise run. Rabbits will not feel safe if they are unable to hide, and a carrier, a few chairs, boxes or rabbit tents may be sufficient.

Excavation in progress. Photo courtesy of Katarina Vallbo, Sweden

Run made of compost fencing. Photo courtesy of Beate Sørland Eltervåg, Norway

> Females dig more than males, and un-neutered females might sneak out to mate with a wild rabbit. Domesticated rabbits in an unsecured run might well get visitors and consequently end up with a surprising litter.

Hutch and run systems

Runaround. Photo courtesy of Runaround.co.uk

The 'Runaround' system (http://www.runaround.co.uk) is a commercially available flexible system of solid or mesh tubes and tunnels that securely fit to wire-mesh runs. They can be set up in any configuration in order to create a secure environment for rabbits, preventing predator attacks, escape, or access to poisonous plants or other hazards. Similar systems can be home-made. They can be adapted to any group of rabbits, and the connecting tunnels can be moved as required.

A rabbit on the run

Escape should be prevented. Rabbits might be hit by a car or scared off by a dog, and they will obviously be defenceless if facing a fox outside the run. However, if not being eaten or chased away, they will normally stay in the neighbourhood. Rabbits do not stray widely if they escape, and may often be found under a porch, in a woodshed, under a bush or any possible imaginable places where there may be room for a rabbit. They will be scared and try to hide, and it is therefore important to look thoroughly and be patient.

If a rabbit has escaped from an enclosure, it is important that this is kept open so that the rabbit has a chance to go home. Rabbits will naturally seek to return to where they can feel safe and get food. However, make sure to secure the run before the rabbits move back in.

It can be challenging to capture rabbits. As prey species, they are very careful and alert, and will not be fooled into a box as would a cat. One should be patient, observe the rabbits, encircling them with compost fences or pens, and get hold of them in a quiet environment instead of chasing them around.

It is possible to use a landing net.

Gaston and Hassan hanging out. Photo courtesy of Katarina Vallbo, Sweden

> Remember that cats are predators. Despite the fact that some rabbits and cats are friends, you must ensure that the rabbits in an outdoor run are never exposed to stranger's cats.
> Cats are everywhere, and they are very insistent hunters that easily climb over fences. They can attack and effectively kill a rabbit, either from hunger or just to play with a prey.

A place to jump and run. Photo courtesy of Marit Emilie Buseth, Norway

Life Outdoors

Harness

There are different harnesses available, and it is important to use a type that does not burden the rabbit's body. Chest Harness and H-harness are the most prevalent types. The Chest Harness can be particularly recommended, as it is equipped with extra straps on the side, making sure the harness is more stable and does not twist around.

If the harness turns around and ends up between the legs, the rabbit will panic, and there are examples of animals that have come loose in such cases.

It is important that the fastening ring is placed on the rearmost strap, to avoid the burden on the neck if the rabbit panics.

One should have a leash that is short enough for one to easily pick up the rabbit when needed. One must be able to protect the prey species against dogs and cats that suddenly show up in the park. Rabbits with long leashes could also twist and panic.

Having a rabbit in a harness is not like walking a dog. Everything should be on the rabbit's terms, and they will often just sniff around in a limited area. Never force the rabbit, but keep in mind that the species will be scared of open places without knowledge of the nearest escape route. To bring a carrier as a possible shelter is therefore recommended.

Mosquitoes

Rabbits get mosquito bites. In nature the rabbits hide from mosquitoes in underground tunnels, and in periods and areas with mosquitoes one must therefore ensure the rabbit's area is protected with mosquito-safe wire mesh and daily go over the ears to check for wounds the rabbit may have scratched to pieces. Such ulcers can easily become infected and should be treated.

Mosquitoes also act as a vector for the spread of myxomatosis, and prevention of bites is an important part of managing the risk of this horrible disease. Products such as permethrin-based insect repellent and killers (e.g. Xenex spot on) may be applied to the rabbit. Note that these products are toxic to cats. Local insect repellents, such as citronella or permethrins, can be used around the rabbit. Insect-repelling plants such as sage, rosemary, thyme, mint, lemon balm, basil, lavender and catnip may be planted, but not all are safe for rabbits to eat.

Mosquitoes like to lay their eggs in stagnant water, so drain or apply airtight covers to any outside water containers. A thin film of vegetable oil on the surface suffocates mosquito larvae. Mosquito-killing bacteria

A sleeping loft in every pen. The rabbits have access to an outdoor run but enjoy staying in their house.
Photo courtesy of Marit Emilie Buseth, Norway

(*Bacillus thuringiensis*; Bt), fish (mosquito fish; *Gambusia affinis*) and other predators (e.g. frogs, toads, dragonflies and damsel flies) also help to decrease mosquito numbers.

Rats and mice

Keep food securely shut away in closed containers to minimize the risk of rodent infestation, disease spread and trauma.

Notes

[1] Harcourt-Brown, F. (2002) *Textbook of Rabbit Medicine*. Butterworth-Heinemann, Oxford, UK.
[2] Hawkins, P. (2008) Refining rabbit care. A resource for those working with rabbits in research. Available at: http://www.rspca.org.uk/servlet/BlobServer?blobtable=RSPCABlob&blobcol=urlblob&blobkey=id&blobwhere=1213709292078&blobheader=application/pdf (accessed 10 October 2010).

Mathias, Mikke, Marikken and Max digging in their garden. Photo courtesy of Marit Emilie Buseth, Norway

Håvard received a second chance. Photo courtesy of Aksel Hunstad, Norway

12

Reproduction and Breeding Control

Abandoned and Unwanted Rabbits

Most rabbits in this book have been lucky, whether they have had responsible caretakers their entire life, or they have been rescued and given a second chance.

It is a hopeless challenge to help the increasing number of homeless and abandoned rabbits, while not at the same time limiting the number of available rabbits. There is a continuous rush of people wanting to dispose of family animals they once claimed responsibility for, and it is important to prevent more individuals being produced.

Anyone can help to improve the situation by providing information about rabbits and their needs and thus help to increase awareness of rabbits, either in the neighbourhood, in the local pet store or through engaging in rescue work. Many of the rabbits in this book have been given a second chance thanks to attentive neighbours who cared.

Despite good intentions, accidents will happen. Rabbits will become pregnant due to confusion about the rabbit's sex, early sexual maturity and lack of knowledge about appropriate precautions, mating through fences or other unfortunate meetings between un-neutered animals. To ensure that the needs of breeding does and newborn rabbits are taken care of, this chapter will provide knowledge on the subject.

I will nevertheless begin with an explanation of why domestic rabbits never can be released into the wild, which many still seem to believe is acceptable.

Wild versus Hutched Rabbits

Observations of wild and domesticated rabbits were analysed[1] as early as 1964. Rabbits born and reared in cages, and rabbits born and reared in pens, were released into enclosures and compared with wild rabbits. Investigations indicated why domesticated rabbits are more susceptible than wild rabbits to the attacks of predators. Despite similar behaviour according to grazing, sociality and activity, the domesticated rabbits seem to rest more above ground during the day. They were thus more exposed and were accordingly more easily taken by predators. In addition, when they had an engineered and unnatural pattern, they were not protected by a camouflage-coloured coat. Also, those reared in cages suffered from injury and disease due to former spatial restriction.

In the wild, rabbits live in colonies, and their survival and wellbeing depends on cooperation and access to burrows, increased defence against predators while grazing, general psychological benefits through mutual grooming, and potential partners and offspring. An abandoned rabbit will survive in neither the forest nor the park.

Breeding Control

I am obviously fond of rabbits, so why would I prevent breeding? It is simply that there are more rabbits than there are good homes, and further additions to the species are strictly speaking not in the rabbit's interest.

Kits are cute, but in a few months they are all grown-ups, they become hormonal and suddenly uncontrollable and mischievous, and when they no longer behave like a docile and cuddly little baby many rabbits are no longer wanted. As a consequence many end up neglected and tucked away animals.

Why not rather take care of someone who already needs a second chance?

Melis also had a second chance. Photo courtesy of Marit Emilie Buseth, Norway

Suitable Rabbits

Nevertheless, if one ends up with a litter, it is important to have knowledge about breeding does, newborn rabbits, and practical and necessary arrangements.

First, one must have knowledge of the mother's and father's genetic origin. One should confirm that the rabbits have no heritable defects, for example so that breeding from animals with poor dentition does not take place.

A female rabbit should also be of an appropriate age. She should be over 6 months and fully developed, yet not older than 2 years. A female rabbit over 2 years will often have an inflexible pelvis and the birth canal will have become too narrow for her to be exposed to a birth. Complications can more easily occur in older females.

Rabbit Ovulation

Rabbits do not have a regular oestrus cycle. They are induced ovulators, and ovulation takes place after stimulation due to mating, or even the proximity of an intact male. In the wild they are reproductively active between April and September, but when the conditions are right, as is the case for our domesticated rabbits, they breed all year round.[2]

With most mammals, ovulation take place at regular intervals, but the rabbit has evolved to give birth to a number of kits and can also be inseminated again directly after giving birth (at the 'post-partum' oestrus).

Pseudopregnancy is the result of ovulation without fertilization, and can lead to pregnancy-like behaviour, including aggression, fur plucking and nest making.

> Female rabbits can conceive immediately after giving birth, so it is important to separate mother and father prior to the birth otherwise the rabbit will probably have a new litter after 30 days.

Feeding the Breeding Doe

For the first 3 weeks of pregnancy the amount and type of food offered should not be changed. The amount of pellets offered should be increased gradually over the last 10 days of pregnancy and the first week of lactation until, at that point, it is approximately double her normal intake. Hay and the like should of course always be offered ad lib.[3]

Housing Conditions and a Rabbit Nest

A rabbit nest. Photo courtesy of Malin Hultman, Sweden

Housing conditions have to be prepared and adapted before birth. In addition to the need to separate the mother and father prior to delivery, one must facilitate and ensure the particular needs of breeding does. With knowledge about breeding behaviour in the wild one can minimize stress for both the mother and her offspring.

A few days before giving birth the doe will find a spot and construct a nest of her own fur and other nesting material, such as straw or hay. Pregnant females will protect their nests and will therefore often behave more aggressively than normal during this period, so leave the soon-to-be mother and her nest alone. The nest box itself should be located outside the mother's usual resting point, to prevent her from sitting on the roof, something that can stress the kits. It is also of major importance that she has the ability to completely withdraw from the tiny rabbits. In the wild, rabbits will typically visit the nest and feed the kits for only about 4 minutes a couple of times every 24 hours, and this absent parenting is a strategy for not attracting unnecessary attention from predators. If not allowed to retreat from her young, the mother may display frustrated behaviour and might even harm her kits and cause infant mortality.[4]

The kits will gradually be more mobile, and after about 2 weeks they will be more often outside the nest exploring. They will probably try to suck milk from their mother more frequently, but she will then escape from them and rest somewhere else.

It is also very important that little ones have room to fool around and develop good motor skills. Good housing conditions are thus essential for both the mother and her kits' wellbeing. A hutch is not enough!

Stress in Breeding Does and Kit Mortality

A breeding doe who either kills or neglects her offspring is a stressed rabbit. Unsatisfactory housing, inability to withdraw from the kits and the presence of other females, even if kept in separate cages, are triggers that may lead to infanticide.[5]

However, it is not normal for rabbits to kill their first born if they are not being exposed to such stressors. Unfortunately, it seems to be a common belief that rabbits have high proportions of kit mortality, probably due to the above-mentioned stressful breeding situations and following fatal results. Wild rabbits are usually good mothers, but it is important to create good environmental conditions for domesticated breeding rabbits to feel safe and comfortable.

Neither handling nor socialization will lead to infanticide. This is another common misconception, which unfortunately leads to many unsocialized and consequently nervous kits. As long as the required adaptations are taken care of, gentle handling of the kits from an early age is recommended to ensure well adapted and secure animals later in life.[6]

Handling the Kits

The effect of early handling and socialization is well known and important in reducing fear of humans later in life.[7] After birth one should check how many kits there are, see if there are any dead and remove these. The litter can be handled

Space enough for the kits to crawl around and the mother to get away. She can also move outside the enclosure. The mother gave birth the day after being rescued. Photo courtesy of Linn Krogstad, Norway

gently, around nursing time, during the first week after birth[8] but should be left mostly alone until they are almost 10 days old. Then it is time to begin to socialize them, still gently. Hold them in your hand for a while, let them get to know people, become used to different smells, normal sounds and movements.

The Kits

Rabbits are born hairless. They are blind and deaf and will lie in a clump together with their siblings. They are very sensitive to cold but stay nice and warm in the nest. If one of the kits is ostracized or falls out of the nest, you have to assist and put it back. When the fur grows out after a few days they will be more resistant.

The eyes open at about 10 days and the hearing probably develops at the same time. They will begin to explore and consume solid food at around 3 weeks of age, but the mother should continue to provide milk as long as possible. Rabbit kits have very sensitive digestion and the mother's milk is important for improving the immune system, although the rabbits eventually will eat more hay and grass. The presence of 'milk oil' from the mother is important in preventing bacterial enteritis at this crucial time. When they start eating solid food, it is important that they always have access to hay and fibrous pellets intended for young and active rabbits.

If the mother is dead or if for some reason does not have breast milk, assisted feeding must be ensured. Please note that one should not give ordinary dairy products. Lactol, Esbilac or goat's milks are often used. In Norway, we may even use Viking Milk. Be aware that a milk substitute does not contain all the bacteria necessary for developing a healthy digestion, so that a probiotic should be provided until the rabbit is about 5 months old to prevent diarrhoea and digestive upsets. If using a probiotic powder, add it to the milk just before feeding the babies, as it will ferment if left in warm milk. Also keep in mind that the milk provided for the kits must maintain the body temperature of the rabbit.

When providing milk to young rabbits, it is important not to force the animal by putting syringes into their mouths. Give milk using a pipette; keep it close to the rabbit's mouth, squeeze out

A pair of rabbits staying outside in their run during summer. Photo courtesy of Nanna Gjerding, Denmark

Tøffen was abandoned and all alone. This picture was taken minutes before we rescued him. For 6 years he has now lived a happy life with his rabbit companion. Photo courtesy of Marit Emilie Buseth, Norway

some milk and let the kit suck as much as it wants. This can be done five times a day or as often as needed. The milk will not be as nutritious as breast milk, and one must therefore provide nutrition more frequently than what the mother would have done.

Time for Relocation

Unfortunately there are many who remove baby rabbits from their mother and siblings and sell them from the age of 5 or 6 weeks old. This is too early, since the kits should be allowed to be socialized as rabbits and enjoy the benefits of staying with the family as long as possible. It is the authors' opinion that kits should not be removed until they are at least 8 weeks of age, and then preferably with a sister or brother.

If the mother and children have well-adapted residential areas, they will have the pleasure of living together.

If a litter is housed together for a long time, as often is the case with rabbits in rescues, the males should be separated from the mother or neutered at the age of 10–11 weeks. The males may become sexually mature from 11–12 weeks and they would otherwise be able to make their mother pregnant again. Females mature later, but intact females and males should not be kept together after they have reached 12–13 weeks. At this time one should make sure to neuter the males, or to separate them until this is done.

It is important to note that kits should receive a great deal of human contact during this period of life, as it has shown that tender and frequent handling of young rabbits reduces fear of humans later in life, making them more adaptive and secure.[9]

Reproduction and Breeding Control

What to consider before breeding rabbits

Rabbits can have many kits, and one can theoretically end up with 12 kits in a litter. It is difficult enough to find good homes for five kits, and any home you might find could rather have helped a rabbit who eagerly awaited a second chance.

- Can you provide adequate housing for ten rabbits over the 10 years ahead? If you do not get the complete litter rehomed, you will be responsible for providing them with good living conditions for the rest of their lives. Different sexes must be neutered or separated by 11 weeks of age, the mother and father must be separated prior to birth, and all have needs for socialization, movement and play.
- Do you have enough room, time and finances to provide a responsible home for ten rabbits?
- Do you know the rabbits you want to breed on? Do you know their genetic history? Are there any dental errors or hereditary diseases? Are they the result of complicated and prolonged inbreeding?
- Is the female of appropriate age? Is she over 6 months but not exceeding 2 years for first-time mothers?
- Do you have the finances to provide them with costly veterinary treatment when needed?
- Do you have necessary knowledge and resources to monitor new homes and make sure that they provide a good life for the rabbits for which you are responsible?

Notes

[1] Stodart, E. and Myers, K. (1964) A comparison of behavior, reproduction, and mortality of wild and domestic rabbits in confined population. *CSIRO Wildlife Research* 9(2), 144–159.

[2] Elliott, S. and Lord, B. (2013) Reproduction. In: *BSAVA Manual of Rabbit Medicine*. BSAVA, Gloucestershire, UK.

[3] Elliott, S. and Lord, B. (2013) Reproduction. In: *BSAVA Manual of Rabbit Medicine*. BSAVA, Gloucestershire, UK.

[4] Hawkins, P. (2008) Refining rabbit care. A resource for those working with rabbits in research. Available at: http://www.rspca.org.uk/servlet/BlobServer?blobtable=RSPCABlob&blobcol=urlblob&blobkey=id&blobwhere=1213709292078&blobheader=application/pdf (accessed 10 October 2010).

[5] Hawkins, P. (2008) Refining rabbit care. A resource for those working with rabbits in research. Available at: http://www.rspca.org.uk/servlet/BlobServer?blobtable=RSPCABlob&blobcol=urlblob&blobkey=id&blobwhere=1213709292078&blobheader=application/pdf (accessed 10 October 2010).

[6] Bilkò, A. and Altbãcker, V. (2000) Regular handling early in the nursing period eliminates fear responses toward humans beings in wild and domestic rabbits. *Developmental Psychobiology* 36(1), 78–87.

[7] Anderson, C.O., Denenberg, V.H. and Zarrow, M.X. (1972) Effects of handling and social isolation upon the rabbit's behaviour. *Behaviour* 43, 165–175.

[8] Csatadi, K. *et al.* (2005) Even minimal human contact linked to nursing reduces fear responses towards humans in rabbits. *Applied Animal Behaviour Science* 95, 123–128.

[9] Jezierski, T.A. and Konecka, A.M. (1995) Handling and rearing results in young rabbits. *Applied Animal Behaviour Science* 46(3), 243–250.

Tøffen and Dina. Photo courtesy of Marit Emilie Buseth, Norway

Melis, a happy house rabbit. Photo courtesy of Marit Emilie Buseth, Norway

Epilogue

The love I feel for my rabbits makes it impossible to tolerate others suffering as a result of poor husbandry. This should be easy to improve, as the majority of caretakers want what is best for the animals for which they are responsible, and because most distress follows lack of knowledge about the species' natural needs. I am therefore grateful to be part of making welfare-oriented knowledge more accessible, and thus improve many rabbits' lives.

In addition to being companion animals, rabbits are a species we happen to put in many different categories. However, the challenge is that rabbits are rabbits, regardless of whether they are classified as companion rabbits, laboratory rabbits, meat rabbits, fur rabbits, exhibition rabbits or whatever category we find useful. All these rabbits are descendants of the European wild rabbit, and share the same physical and behavioural needs. However, we treat them differently, and there are even different laws and regulations based on what 'type' of rabbit it is.

Those of us who care about rabbits should be aware of our responsibility for all rabbits and avoid products that contain angora, fur or rabbit meat. The undercover video, shot by PETA Asia in 2013, revealed routine cruelty to angora rabbits, whose long, soft fur is often used in sweaters and accessories. The barbaric ordeal shocked the whole world, and one could see workers violently ripping the fur from the animals' sensitive skin as they screamed in pain. Disclosures relating to fur and meat production happen continuously around the world.

I assume that most people who have read *Rabbit Behaviour, Health and Care* have an interest in and care for rabbits. We know rabbits. We are aware that they have emotions, that they are capable of learning, and that they have a wide spectrum of needs that must be fulfilled to achieve their welfare. However, all animals have needs – cattle, monkeys, fish, pigs and sheep have the ability to learn and experience pleasure and suffering.

How can we treat as many animals with so little consideration when we simultaneously endeavour to give our dog the most satisfying life? Why is it okay for so many people that a hen must stay a lifetime in a malodorous and dark cage the size of an A4 sheet of paper, whilst at the same time they will ensure that their own cat can be comfortable in the sun?

Photo courtesy of Anton Kragh, Norway

Photo courtesy of Anton Kragh, Norway

How can we accept that a pig will never run outdoors or have the opportunity to practise normal behaviour? Many sows are restricted from any movement when being kept in gestation stalls due to being pregnant most of the time.[1]

How can we support separating the offspring of dairy cows from their mother immediately after birth, to be fed milk substitute from a bucket?

How do we choose which animals are worthy of our respect? A pig is for example as sensitive as a dog, but most people would be upset if a golden retriever and cocker spaniel had to live under conditions that are normal in most barns. Many would likewise find that it was ethically reprehensible if kittens were slaughtered in the same way as chickens.

Today's agriculture cannot be compared with earlier production, as modern farming is all about keeping the cost down and productivity up. It is our responsibility to choose what we want to support, and if consumers of meat, eggs and dairy products show commitment and moderation, one will eventually affect the poor living conditions of many animals in industrial agriculture. Choose plant-based alternatives or select products from farms that focus on improving animal welfare.

Marit Emilie Buseth
Oslo, Norway, 6 November 2014

The question is not, "Can they reason?" nor, "Can they talk?" but "Can they suffer?"
Jeremy Bentham

Note

[1] DeMello, M. (2012) The making and consumption of meat. In: *Animals and Society. An Introduction to Human-Animal Studies*. Colombia University Press, New York, pp. 126–146.

Photo courtesy of Andy Purivance, USA

Wild rabbit eating. Photo courtesy of Joshua Davis (http://joshuadavisphotography.com)

Bibliography

Anderson, C.O., Denenberg, V.H and Zarrow, M.X. (1972) Effects of handling and social isolation upon the rabbits behaviour. *Behaviour* 43, 165–175.

Batchelor, G.R. (1999) The laboratory rabbit. In: Poole, T. and English, P. (eds) *The UFAW Handbook on the Care and Management of Laboratory Animals*, 7th edn. Blackwell Science, Oxford, UK, pp. 395–408.

Bell, D. (2008) Understanding wild rabbit behavior. *Rabbiting On*, winter, 10–11.

Berthelsen, H. and Hansen, L.T. (1999) The effect of hay on the behaviour of caged rabbit (*Oryctolagus cuniculus*). *Animal Welfare* 8, 149–157.

Berthelsen, H. and Hansen, L.T. (2000) The effect of environmental enrichment on the behavior of caged rabbits (*Oryctolagus cuniculus*). *Applied Animal Behaviour Science* 68, 163–178.

Bilkò, A. and Altbâcker, V. (2000) Regular handling early in the nursing period eliminates fear responses toward humans beings in wild and domestic rabbits. *Developmental Psychobiology* 36(1), 78–87.

Boers, K., Gray, G., Love, J., Mahmutovic, Z., McCormick, S., Turcotte, N. and Zhang, Y. (2002) Comfortable Quarters for Rabbits in Research Institutions. Available at: http://www.awionline.org/pubs/cq02/Cq-rabbits.html (accessed 10 October 2011).

Bouvier, A.C. and Jacquinet, C. (2008) Pheromone in rabbits: Preliminary technical results on farm use in France. Available at: http://world-rabbit-science.com/WRSA-Proceedings/Congress-2008-Verona/Papers/R-Bouvier.pdf (accessed 8 April 2013).

Bracht, P.B. (2009) Medisinsk og kirurgisk behandling av gnagere, kanin og ilder. *Proceedings of etterutdanning fra Norges Veterinærhøyskole*, Oslo, Norway.

Bradley, B.T. (2006) Rabbit behavior. In: Bradley Bays, T., Lightfoot, T.L. and Mayer, J. (eds) *Exotic Pet Behavior. Birds, Reptiles, and Small Mammals*. Saunders Elsevier, St Louis, Missouri, pp. 1–49.

Brodbelt, D.C. (2006) The confidential enquiry into perioperative small animal fatalities. Available at: http://www.rvc.ac.uk/Staff/Documents/dbrodbelt_thesis.pdf (accessed 10 February 2013).

Broom, D.M. (2010) Confinement. In: Mills, D.S., Marchant-Forde, J.N., McGreevy, P.D., Morton, D.B., Nicol, C.J., Phillips, C.J.C., Sandoe, P. and Swaisgood, R.R. (eds) *The Encyclopedia of Applied Animal Behaviour and Welfare*. CAB International, Wallingford, UK, pp. 125–127.

Buijsa, S., Keeling, L.J., Vangestelc, C., Baertd, J., Vangeytee, J. and Tuyttens, F.A.M. (2011) Assessing attraction or avoidance between rabbits: comparison of distance-based methods to analyse spatial distribution. *Animal Behaviour* 82, 1235–1243.

Buijsa, S., Keeling, L.J. and Tuyttens, F.A.M. (2011) Behaviour and use of space in fattening rabbits as influenced by cage size and enrichment. *Applied Animal Behaviour Science* 134, 229–238.

Burghardt, G.M. (2005) *The Genesis of Animal Play*. MIT Press, Cambridge, Massachusetts.

Calvete, C., Estrada, R., Lucientes, J.J. and Villafuerte, O.R. (2004) Effects of vaccination against viral haemorrhagic disease and myxomatosis on long-term mortality rates of European wild rabbits. *The Veterinary Record* 155, 388–392.

Chapman, J.A. and Flux, J.E.C. (eds) (1990) *Rabbits, Hares and Pikas: Status Survey and Conservation Action Plan*. IUCN, Gland, Switzerland.

Chu, L., Garner, J. and Mench, J.A. (2003) A behavioral comparison of New Zealand White rabbits (*Oryctolagus cuniculus*) housed individually or in pairs in conventional laboratory cages.

Applied Animal Behaviour Science 85, 121–139.

Clauss, M. (2012) Clinical technique: feeding hay to rabbits and rodents. *Journal of Exotic Pet Medicine* 21, 80–86.

Cousquer, G. (2007) Digestion in rabbits. How it works. *Rabbiting On*, winter, 14–17.

Cruise, L.J. and Brewer, N.R. (1994) Anatomy. In: Mannings, P.J., Ringler, D.H. and Newcomer, C.E. (eds) *The Biology of the Laboratory Rabbit*, 2nd edn. Academic Press, San Diego, California, pp. 47–61.

Davis, S.E. and DeMello, M. (2003) *Stories Rabbits Tell. A natural and cultural history of a misunderstood creature.* Lantern Books, New York.

Dawkins, M.S. (1988) Behavioural deprivation: a central problem in animal welfare. *Applied Animal Welfare Science* 20, 209–225.

Delibes-Mateos, M., Ferraras, P. and Villafuerte, R. (2006) Rabbit populations and game management: the situation after 15 years of rabbit haemorrhagic disease in central-southern Spain. Available at: http://www.federaciongalegadecaza.com/biblioteca/coello/CIENTIFICAS_061.pdf (accessed 10 July 2012).

Delibes-Mateos, M., Ferreras, P. and Villafuerte, R. (2009) European rabbit population trends and associated factors: a review of the situation in the Iberian Peninsula. *Mammal Review* 39(2), 124–140.

DeMello, M. (2012) Animal behaviour studies and ethology. In: *Animals and Society. An Introduction to Human-Animal Studies.* Colombia University Press, New York, pp. 349–373.

DeMello, M. (2012) The making and consumption of meat. In: *Animals and Society. An introduction to Human-Animal Studies.* Colombia University Press, New York, pp. 126–146.

Dennis, R. (2010) Pheromone. In: Mills, D.S., Marchant-Forde, J.N., McGreevy, P.D., Morton, D.B., Nicol, C.J., Phillips, C.J.C., Sandoe, P. and Swaisgood, R.R. (eds) *The Encyclopedia of Applied Animal Behaviour and Welfare.* CAB International, Wallingford, UK, pp. 468–470.

Dixon, L.M., Hardiman, J.R. and Cooper, J.J. (2010) The effects of spatial restriction on the behaviour of rabbits (*Oryctolagus cuniculus*). *Journal of Veterinary Behaviour: Clinical Applications and Research* 5(6), 302–308.

Drescher, B. and Loeffler, K. (1991) Einfluβ unterschiedlicher Haltungsverfahren und Bewegunsmöglichkeiten auf die Kompakta der Röhrenknochen von Versuchs- und Fleischkaninchen. *Tierärztliche Umschau* 46, 736–741.

Dykes, L. (2006) Problem behavior in rabbits. Dr Linda Dykes reports on a talk given by lecturer in animal behavior, Dr Anne McBride. *Rabbiting On*, summer, 14–15.

Dykes, L. (2006) Problem behavior in rabbits – Part 2. Dr Linda Dykes reports on a talk giving by Dr Anne McBride. *Rabbiting On*, autumn, 8–10.

Edgar, J.L. and Mullan, S.M. (2011) Knowledge and attitudes of 52 UK pet rabbit owners at the point of sale. *Veterinary Record* 168(13), 353.

Edwards, L.N. (2010) Animal well-being and behavioural needs on the farm. In: Granding, T. (ed.) *Improving Animal Welfare. A Practical Approach.* CAB International, Wallingford, UK, pp. 139–159.

Elliott, S. and Lord, B. (2013) Reproduction. In: *BSAVA Manual of Rabbit Medicine.* BSAVA, Gloucestershire, UK (in press).

Fairham, J. and Harcourt-Brown, F. (1999) Preliminary investigation of the vitamin D status of pet rabbits. *The Veterinary Record* 145, 452–454.

Francis, B. (2012) Back to nature. Wild vs domestic diet. *Rabbiting On*, Autumn, 12–13.

Frantz, R., Kreuzer, M., Hummel, J., Hatt, J.M. and Clauss, M. (2011) Differences in feeding selectivity, digesta retention, digestion and gut fill between rabbits (*Oryctolagus cuniculus*) and guinea pigs (*Cava porcellus*). Available at: http://www.zora.uzh.ch/49675/5/RF_RabbitsGPigsPhysio_revision.pdf (accessed 20 December 2012).

Fraser, D. and Nicol, C.J. (2011) Preference and motivation rersearch. In: Appleby, M.C., Mench, J.A., Olsson, I.A.S. and Hughes, B.O. (eds) *Animal Welfare*, 2nd edn. CAB International, Wallingford, UK, pp. 183–199.

Fraser, M.A. and Girling, S.J. (2009) *Rabbit Medicine and Surgery for Veterinary Nurses.* Wiley-Blackwell, Chichester, UK.

Girling, S. (2009) *Veterinary Nursing of Exotic Pets.* Blackwell Publishing, Oxford, UK.

Gunn, D. and Morton, D.B. (1993) The behavior of single-caged and group-housed laboratory rabbits. In: *Proceedings of the Fifth Federation of European Laboratory Animal Science Association (FELASA)* Symposium, 80–84.

Gunn, D. and Morton, D.B. (1995) Inventory of the behaviour of New Zealand White rabbits in laboratory cages. *Applied Animal Behaviour Science* 45, 277–292.

Gunn, D. and Morton, D.B. (1998) Rabbits. Available at: http://www.nal.usda.gov/awic/pubs/enrich/rabbits.htm (accessed 8 April 2013).

Gunn-Dore, D. (1997) Comfortable quarters for laboratory rabbits. Available at: http://www.awionline.org/pubs/cq/five.pdf (accessed 1 July 2012).

Hawkins, P., Hubrecht, R., Buckwell, A., Cubitt, S., Howard, B., Jackson, A. and Poirier, G.M. Refining rabbit care. A resource for those working with rabbits in research. Available at: http://www.rspca.org.uk/servlet/BlobServer?blobtable=RSPCABlob&blobcol=urlblob&blobkey=id&blobwhere=1213709292078&blobheader=application/pdf (accessed 10 October 2011).

Harcourt-Brown, F. (2002) Dental disease. In: *Textbook of Rabbit Medicine*. Butterworth-Heinemann, Oxford, UK, pp. 165–205.

Harcourt-Brown, F. (2002) Digestive disorders. In: *Textbook of Rabbit Medicine*. Butterworth-Heinemann, Oxford, UK, pp. 249–291.

Harcourt-Brown, F. (2002) *Textbook of Rabbit Medicine*. Butterworth-Heinemann, Oxford, UK.

Harcourt-Brown, F. (2005) Dental disease. Chronic illness in rabbits. *Rabbiting On*, Summer, 15–17.

Harcourt-Brown, F. (2009) Dental disease in pet rabbits: 1. Normal dentition, pathogenesis and aetiology. *In Practice* 31, 370–379.

Harcourt-Brown, F. and Fairham, J. (1999) Preliminary investigation of the vitamin D status of pet rabbits. *The Veterinary Record* 145, 452–454.

Healtley, J.J. and Smith, A.N. (2004) Spontaneous neoplasms of lagomorphs. *Veterinary Clinics of North America: Exotic Animal Practice* 7, 561–577.

Hedley, J. (2011) Critical care of the rabbit. *In Practice* 33, 386–391.

Held, S.D.E. (1996) Group-Housing of Female Laboratory Rabbits – Studies on Behaviour and Immunocompetence. PhD dissertation, University of Wales, Aberystwyth, UK.

Held, S.D.E., Turner, R.J. and Wooton, R.J. (1995) Choices of laboratory rabbits for individual or group-housing. *Applied Animal Behaviour Science* 46(1), 81–91.

Hilyer, E.V. (1994) Pet rabbits. *Veterinary Clinics of North-America: Small Animal Practice* 24, 25–69.

House Rabbit Society. FAQ: Litter training. Available at: http://www.rabbit.org/faq/sections/litter.html (accessed 20 February 2012).

Huang, C., Mi, M. and Vogt, D. (1981) Mandibular prognathism in the rabbit: discrimination between single-locus and multifactoral models of inheritance. *Journal of Heredity* 72(4), 296–298.

Hull, W.L., Brooks, D.L. and Bean-Knutdsen, D. (1991) Response of adult New Zealand white rabbits to enrichment objects and paired housing. *Laboratory Animal Science* 41(6), 609–612.

Jezierski, T.A. and Konecka, A.M. (1995) Handling and rearing results in young rabbits. *Applied Animal Behaviour Science* 46(3), 243–250.

Johnson, J.H. and Wolf, A.M. (1993) Ovarian abscesses and pyometra in domestic rabbit. *JAVMA* 5, 667–669.

Keeble, E. (2011) Encephalitozoonosis in rabbits – what we do and don't know. *In Practice* 33, 426–435.

Keeling, L.J., Rushen, J. and Duncan, J.H. (2011) Understanding animal welfare. In: Appleby, M.C., Mench, J.A., Olsson, I.A.S. and Hughes, B.O. (eds) *Animal Welfare*, 2nd edn. CAB International, Wallingford, UK, pp. 13–25.

Kersten, A.M.P., Meijsser, F.M. and Metz, J.H.M. (1989) Effects of early handling on later open-field behavior in rabbits. *Applied Animal Behaviour Science* 23(2), 157–167.

Krohn, T.C., Ritskes-Hoitinga, J. and Svendsen, P. (1999) The effects of feeding and housing on the behaviour of the laboratory rabbit. *Laboratory Animals* 33, 101–107.

Law, G. (2010) Animal enclosure. In: Mills, D.S., Marchant-Forde, J.N., McGreevy, P.D., Morton, D.B., Nicol, C.J., Phillips, C.J.C., Sandoe, P. and Swaisgood, R.R. (eds) *The Encyclopedia of Applied Animal Behaviour and Welfare*. CAB International, Wallingford, UK, pp. 18–20.

Lennox, A.M. (2008) Clinical technique: small exotic companion mammal dentistry – anesthetic considerations. *Journal of Exotic Pet Medicine* 17(2), 102–106.

Lidfors, L. (1997) Behavioural effects of environmental enrichment for individually caged rabbits. *Applied Animal Behaviour Science* 52, 157–169.

Lindsey, J.R. and Fox, R.R. (1994) Inherited diseases and variations. In: Manning, P.J., Ringler, D.H. and Newcomer, C.E. (eds) *The Biology of the Laboratory Rabbit*, 2nd edn. Academic Press, London, pp. 293–320.

Lockley, R.M. (1965) *The Private Life of the Rabbit*. A.Wheaton & Co., Exeter, UK.

Loeffler, K., Drescher, B. and Schulze, G. (1991) Einfluβ unterschiedlicher Haltunsverfahren auf das Verhalten von Versuchs- und Fleischkaninchen. *Tierärztliche Umschau* 46, 471–478.

Longley, L. (2009) An update on anaesthesia and post-anaesthetic care. *Proceedings of the 2009 RWF Conference*, Veterinary Professional Notes, 28 February 2009, Nottingham, UK.

Lord, B. (2010) Rabbit anaesthesia – how we can make it safer. *Proceedings of the 2010 RWF Conference*, Veterinary Professional Notes, 22 May 2010, Manchester, UK.

Love, J.A. (2009) Group housing: meeting the physical and social needs of laboratory rabbits. *Laboratory Animal Science* 44(1), 5–11.

Mancinelli, E., Keeble, E., Richardson, J. and Hedley, J. (2014) Husbandry risk factors associated with hock pododermatitis in UK pet rabbits (*Oryctolagus cuniculus*). *Veterinary Record* 174(17), 429.

Marchant-Forde, J. (2010) Aggression. In: Mills, D.S., Marchant-Forde, J.N., McGreevy, P.D., Morton, D.B., Nicol, C.J., Phillips, C.J.C., Sandoe, P. and Swaisgood, R.R. (eds) *The Encyclopedia of Applied Animal Behaviour and Welfare*. CAB International, Wallingford, UK, pp. 8–10.

Martorell, J. (2014) Scoring pododermatitis in pet rabbits. *Veterinary Record* 174, 427–428.

Mason, G.J. and Burn, C.C. (2011) Behavioural restriction. In: Appleby, M.C., Mench, J.A., Olsson, I.A.S. and Hughes, B.O. (eds) *Animal Welfare*, 2nd edn. CAB International, Wallingford, UK, pp. 98–119.

McBride, A. (1988) *Rabbits and Hares*. Whittet Books, Suffolk, UK.

McBride, A. (1998) *Why Does my Rabbit?* Souvenir Press, London.

McBride, A. (2009) Help for nervous and aggressive rabbits. *Rabbiting On*, Autumn, 22–25.

McBride, A. (2009) Rampaging rabbits: minimising stress in the rescue centre and veterinary surgery. *Proceedings of the 2009 RWF Conference*, Veterinary Professional Notes, 28 February 2009, Nottingham, UK.

McBride, A., Day, S., McAdie, T., Meredith, A., Hickman, J. and Lawes, L. Trancing rabbits: relaxed hypnosis or a state of fear? Available at: http://www.hopperhome.com/Trancing%20Rabbits-Tonic%20Immobility%20.pdf (accessed 20 May 2010).

Melfi, V. (2010) Enrichment. In: Mills, D.S., Marchant-Forde, J.N., McGreevy, P.D., Morton, D.B., Nicol, C.J., Phillips, C.J.C., Sandoe, P. and Swaisgood, R.R. (eds) *The Encyclopedia of Applied Animal Behaviour and Welfare*. CAB International, Wallingford, UK, pp. 221–223.

Meredith, A. (2005) Clinical and behavioural assessment of tonic immobility response in rabbits. DipZooMed thesis, University of Edinburgh, Scotland, UK.

Meredith, A.M (2010) Animal Welfare Concerns for Companion and Laboratory Rabbits. Information Resources in the Care and Welfare of Rabbits. Available at: http://www.nal.usda gov/awic/pubs/Rabbits/meredith.htm (accessed 12 October 2009).

Meredith, A. (2010) The rabbit GI tract – on a knife edge. *Proceedings of the 2010 RWF Conference*, Veterinary Professional Notes, 22 May 2010, Manchester, UK.

Meredith, A. (2012) Is obesity a problem in pet rabbits? *Veterinary Record* 171, 192–193.

Mills, D.S. (2010) Abandoned animals. In: Mills, D.S., Marchant-Forde, J.N., McGreevy, P.D., Morton, D.B., Nicol, C.J., Phillips, C.J.C., Sandoe, P. and Swaisgood, R.R. (eds) *The Encyclopedia of Applied Animal Behaviour and Welfare*. CAB International, Wallingford, UK, pp. 8–10.

Moore, L. and Smith, K. (2008) *When your Rabbit Needs Special Care. Traditional and Alternative Healing Methods*. Santa Monica Press, Santa Monica, California.

Morrell, J.M. (1989) Hydrometra in the rabbit. *Veterinary Record* 16, 325.

Mullan, S.M. and Main, D.C. (2006) Survey of the husbandry, health and welfare of 102 pet rabbits. *Veterinary Record* 159, 103–109.

Nelissen, M. (1975) On the diurnal rhythm of activity of *Oryctolagus cuniculus* (Linne, 1758). *Acta Zool Pathol Antwerp* 61, 3–18.

Oglesbee, B.L. (2006) *The 5-Minute Veterinary Consult, Ferret and Rabbit*. Blackwell Publishing, Oxford, UK.

PDSA (2012) PDSA Animal Wellbeing Report. Available at: https://www.pdsa.org.uk/pet-health-advice/pdsa-animal-wellbeing-report (accessed 9 November 2012).

Pellis, S.M. and Pellis, V.C. (2010) Play. In: Mills, D.S., Marchant-Forde, J.N., McGreevy, P.D., Morton, D.B., Nicol, C.J., Phillips, C.J.C., Sandoe, P. and Swaisgood, R.R. (eds) *The Encyclopedia of Applied Animal Behaviour and Welfare*. CAB International, Wallingford, UK, pp. 477–479.

Pizzi, R. (2009) Osteoarthritis in rabbits. *Rabbiting On*, Winter, 7–9.

Redrobe, S. (2011) Is a hutch enough? A comparison between hutch only, hutch & pen and hutch & runaround systems. *Proceedings from The 2011 RWF Conference*, Solihull, UK, 29 October 2011.

Reusch, B. (2006) Urogenital system and disorders. In: *Rabbit Medicine and Surgery*, 82nd edn. BSAVA Publications, Gloucester, UK.

Romain, P. (2011) The effect of viral haemorrhagic disease (vHD) vaccination on the longevity of pet rabbits (*Oryctolagus cuniculus*) in Scotland. In: Roberts, V. (ed.) *British Veterinary Zoological Society Proceedings*, Cheshire, UK, 12–13 November 2011, 44.

Rushen, J. et al. (2008) *The Welfare of Cattle*. Springer, Dordrecht, the Netherlands, 303 pp.

Saito, K., Nakanishi, M. and Hasegawa, A. (2002) Uterine disorders diagnosed by ventrotomy in 47 rabbits. *Journal of Veterinary Medical Science* 64(6), 495–497.

Saunders, R. (2009) Rabbits lumps and bumps. When to worry, what to do! *Proceedings of the 2009 RWF Conference*, Veterinary, 28 February 2009, Nottingham, UK.

Saunders, R. (2010) Urinary tract problems. *Rabbiting On*. 01, 4–5.

Saunders, R.A. and Rees Davies, R. (2005) *Rabbit Internal Medicine*. Blackwell Publishing, Oxford, UK.

Saunders, R. and Whitlock, E. (2012) Nursing hospitalized patients. In: *BSAVA Manual of Exotic Pet and Wildlife Nursing*. British Small Animal Veterinary Association, Gloucester, UK, pp. 129–166.

Sayers, I. (2010) Approach to preventive health care and welfare in rabbits. *In Practice* 32, 190–198.

Schepers, F., Koene, P. and Beerda, B. (2009) Welfare assessment in pet rabbits. *Animal Welfare* 18, 477–485.

Seaman, S. (2002) Laboratory rabbit housing: An investigation of the social and physical environment. Available at: http://www.ufaw.org.uk/pdf/phhsc-schol1-summary.pdf (accessed 15 February 2013).

Seaman, S. (2008) Getting the message. *Rabbiting On*, Winter, 2–7.

Sherwin, C.M. et al. (2004) Studies on the motivation for burrowing by laboratory mice. *Applied Animal Behaviour Science* 88, 343–358.

Spencer, R. (2011) Guidelines for entry into meat rabbit production. Available at: http://www.aces.edu/pubs/docs/U/UNP-0080 (accessed 12 January 2013).

Spinka, M. and Wemelsfelder, F. (2011) Environmental challenge and animal agency. In: Appleby, M.C., Mench, J.A., Olsson, I.A.S. and Hughes, B.O. (eds) *Animal Welfare*, 2nd edn. CAB International, Wallingford, UK, pp. 27–43.

State Government of Victoria (2010) Rabbits and their impact. Available at: http://www.dpi.vic.gov.au/agriculture/pests-diseases-and-weeds/pest-animals/lc0298-rabbits-and-their-impact (accessed 20 January 2012).

Statham, P. (2010) Behavioural need. In: Mills, D.S., Marchant-Forde, J.N., McGreevy, P.D., Morton, D.B., Nicol, C.J., Phillips, C.J.C., Sandoe, P. and Swaisgood, R.R. (eds) *The Encyclopedia of Applied Animal Behaviour and Welfare*. CAB International, Wallingford, UK, pp. 52–53.

Stodart, E. and Myers, K. (1964) A comparison of behavior, reproduction, and mortality of wild and domestic rabbits in confined population. *CSIRO Wildlife Research* 9(2), 144–159.

Tettamanti, M. and Veeraraghavan, P. The impact of facts from the rehabilitation of laboratory rabbits on reliability and evaluation of experimental data. Available at: http://www.icare-worldwide.org/images/poster_4_rabbits.pdf (accessed 2 November 2012).

The Animal Welfare Act (2006) Available at: http://archive.defra.gov.uk/foodfarm/farmanimal/welfare/act/documents/aw-act-2006-memo-101220.pdf (accessed 25 February 2013).

Tschudin, A., Clauss, M., Codron, D., Liesegang, A. and Hatt, J.M. (2011) Water intake in domestic rabbits (*Oryctolagus cuniculus*) from open dishes and nipple drinkers under different water and feeding regimes. *Journal of Animal Physiology and Animal Nutrition* 95, 499–511.

Vastrade, F. (1986) The social behavior of free-ranging domestic rabbits (*Oryctolagus cuniculus*). *Applied Animal Behaviour Science* 16, 165–177.

Vastrade, F.M. (1987) Spacing behaviour of free-ranging domestic rabbits, *Oryctolagus cuniculus*. *Applied Animal Behaviour Science* 18(2), 185–195.

Vinuela-Fernandez, I., Weary, D.M. and Flecknell, P. (2011) Pain. In: Appleby, M.C., Mench, J.A., Olsson, I.A.S. and Hughes, B.O. (eds) *Animal*

Welfare, 2nd edn. CAB International, Wallingford, UK, pp. 64–77.

Virgòs, E., Cabezas-Diaz, S. and Lozano, J. (2007) Is the wild rabbit (*Oryctolagus cuniculus*) a threatened species in Spain? Sociological constraints in the conservation of species. *Biodiversity and Conservation* 16, 3489–3504. Available at: http://www.escet.urjc.es/biodiversos/publica/Virgos_et_al_2007_Biodivers_Conserv.pdf (accessed 8 April 2013).

Von Holst, D., Hutzelmeyer, P., Kaetzke, P., Khaschei, M., Heiko, G. and Schrutka, R.H. (2001) Social rank, fecundity and lifetime reproductive success in wild European rabbits (*Oryctolagus cuniculus*). *Behavioral Ecology and Sociobiology* 51, 245–254.

Walshaw, S. (2006) Euthanasia. In: Meredit, A. and Flecnell, P. (eds) *BSAVA Manual of Rabbit Medicine and Surgery*, 2nd edn. BSAVA Publishing, Gloucester, UK.

Warner, J. (2011) Pets may reduce children's allergy risk. Children who had a dog or cat as infants less likely to become allergic. Available at: http://www.webmd.com/allergies/news/20110613/pets-may-reduce-childrens-allergy-risk (accessed 10 March 2013).

Weisbroth, S.H. (1994) Neoplastic diseases: tumors of the mammary gland. In: *The Biology of the Laboratory Rabbit*, 2nd edn. Academic Press, New York, pp. 345–347.

Wenger, S. (2008) The important role of pain therapy in rabbit medicine. *Rabbiting On*, Summer, 4–7.

Whary, M. (2007) The effects of group housing on the research use of the laboratory rabbit. *Laboratory Animals* 27, 330–341.

Woods, V. (2010) Enrichment. In: Mills, D.S., Marchant-Forde, J.N., McGreevy, P.D., Morton, D.B., Nicol, C.J., Phillips, C.J.C., Sandoe, P. and Swaisgood, R.R. (eds) *The Encyclopedia of Applied Animal Behaviour and Welfare*. CAB International, Wallingford, UK, pp. 221–223.

Wuebel, H. (2010) Sterotypies. In: Mills, D.S., Marchant-Forde, J.N., McGreevy, P.D., Morton, D.B., Nicol, C.J., Phillips, C.J.C., Sandoe, P. and Swaisgood, R.R. (eds) *The Encyclopedia of Applied Animal Behaviour and Welfare*. CAB International, Wallingford, UK, pp. 575–578.

Zoltan, P. (2007) Behaviour of growing rabbits under various housing conditions. *Applied Animal Behaviour Science* 111(3), 342–356.

Index

aggression 31–34, 37–38
    attacking owner's hands 31
    causes 32
    comical dance 32
    definition 31
    fear-related 33–34
    frustration 34
    learning 34, 37–38
    neglected rabbits 35–36
    pain 32
    play activity 32
    redirected 34, 36
    resource-related 32–33
American 'rock rabbits' 7
Animal Welfare Act 21, 120, 162

Bayliscascaris procyonis 110
behaviour 39–51
    abandonment 30
    activation and play 50
    aggressive *see* aggression
    captive 40–42
    fear-related 39
    housing 39–40
    jumping 50–51
    nature and nurture 30–31
    needs and welfare 44–46
    outside activity 50
    play 47–50
    rescue operations 45
    stress 46–47
    and wellbeing 164–165
Belgian Hares 2
bonding process 59–64
    behaviours and needs 62
    caretakers 62
    description 59–60
    EC 62
    fighting 63–64
    hiding places 61
    humping 60
    indoors 61
    litter trays 61
    neutral place 60–61
    stress 61
    supervising 61
    transport 60
breeding 201–203, 205
    control 201–202
    feeding 202
    housing conditions and rabbit nest 202–203
    and kit 203
    stress 203
    time for relocation 205

cages and space 171–173
    customized rabbit room 174
    design 171
    dog cage 173
    exercise yard 173
    freedom of movement 173
    Hol table, IKEA 172
    module-based storage system 173
    requirements 172
    RWAF 171
    Scandinavian design 172
cancer 107–108
    internal masses 107–108
    neutering females 140
    prostatic and testicular 142
    skin neoplasia 107
    treatment options 108
    uterine adenocarcinoma 107
care 20–23
    annual care 23
    daily care 21–22
    human interaction 20
    monthly care 22–23
    weekly care 22
check-up 89–92
    abdomen 91–92
    ears 90
    eyes 90
    healthy/symptoms of illness 89
    heart and lungs 91
    medications/anaesthesia 89

check-up (*continued*)
    nose 90
    rabbit's living conditions 89
    skin 91
    teeth 90–91
    veterinarian 89
cheek teeth 129–131
    examination and treatment 130–131
    herbivores 129
    and incisors 124, 129
    jaw movements 129
    malocclusion and dental disorders 129–130
chin gland lesions 142
cleanliness and hygiene 151–158
    age-related questions 152–153
    communication 153
    droppings 153
    litter tray 153–155
    natural way 151–152
    neutering 152–153
    rabbit habit 156–158
communication 42–44
    chemo-signalling 43
    disapproving rabbit 45
    lavatory 44
    ownership and presence 44
    pheromones 43
    sound 44
    visual expression 44
companion animal 15–27
    allergies 17–18
    'babysitting' 20
    Britain 15–16
    care 20–23
    characteristics 18
    and children 20–21
    costs 18–19
    living conditions 19
    movement opportunities 20
    nutrition 19
    PDSA's PAW Report 16
    rabbit enthusiasts 15
    rabbit selection 23–25
    rehoming and rescues 25–27
    social rabbit 20
    time perspective 17
coprophagy 87, 132
customs and practice 53–54

Den Store Kaninboka 193
dental disease 125–131
    abrasive nature of diet 126
    aradicular 126
    balanced diet 127
    cheek teeth 129–131
    congenital defects/genetic predisposition 128
    dacryocystitis 126
    digestive disorders 126
    epiphora 126
    grasses shape and plant fibre content 126
    grinding 126
    herbivores 126
    incisors 128–129
    infections 126
    malocclusion 128
    mandible 125
    muesli versus pellets 120
    myiasis (flystrike) 126
    nutrition 126
    premolars and molars 125
    symptoms 126
    trauma 128
    tumours/infection 128
    unrestricted access 126
    wounds 126
diarrhoea 97, 101
    coccidiosis infection 97
    digestive system 97
    parasitic gut problems 97
    symptoms of infection 97
    treatment 97
digestive physiology 131–137
    abdominal cavity 132
    abnormalities 131
    balanced diet 131
    bloated stomach 132
    caecotrophs 132
    caecum 131
    constant movement 131
    coprophagy 132
    enzymes and acid 131
    faecal pellets 132
    faeces produced 131
    fibrous low-carb diet 131–134
    GI tract 131
    intestinal obstruction 135–136
    non-obstructive and obstructive ileus 136–137

ectoparasites 100
Encephalitozoon cuniculi (EC) 62, 105–107
enrichments 166–169
    cognitive 169
    Cottontail Cottage 168
    definition 166
    environmental improvements 166
    food-based 168
    internal accomodation 168
    physical 168
    rescue case study 167
    sensory 168

social 168–169
tunnel 168
European rabbit *(Oryctolagus cuniculis)* 5–7

flystrike/fly larvae attacks (myiasis) 101–102
    good diet 102
    hindquarters and damp surfaces, soiling 101
    Lucilia sericata 101
    maggots 101
    moisture 101
    obesity 122
    prevention and treatment 101–102
    regular observation 101

gastrointestinal (GI) stasis 98, 134–135, 137
    blockages 98
    caecal fermentation patterns 134
    complex and sensitive 135
    exercise and movement 137
    fibre and probiotics 137
    fluid therapy 137
    food intake and faeces output 98
    gas reduction drugs 137
    motility drugs 137
    muscular contractions 134
    pain relief 137
    slowing/cessation 98
    triggers 135

habit 156–158
    bathroom 156
    house-clean rabbit 156–157
    litter tray in cage 157–158
    living space arrangement 156
    smell 158
    urinary incontinence 156
hay 118–119
    hay selection 118
    providing hay 118–119
    variety and selective feeding 118
The House Rabbit Society 3
housing and conditions 161–175
    air quality 174
    behaviour and wellbeing 164–165
    cages and space 171–173
    conventional cages 161
    enclosures 161, 170–171
    enrichments 166–169
    environments 161
    free-range housing 161
    lifespan 161
    living environment, European 162
    physical wellbeing 162–164
    rabbit interior 165–166
    reproduction and breeding control 202–203

incisors 128–129
    and cheek teeth 124
    chewing 128
    dental enamel 128
    examination and treatment 129
    malocclusion and dental disorders 128–129
    and molars 95
    small teeth 128

kit 203–205
    feeding 204
    handling 203–204
    milk to young rabbits 204–205
    mortality 203

Lagomorphs 7–13
    Leporidae family (rabbits and hares) 8–10
    Lepus (hares) 10–13
    Ochotonidae family (pikas) 7–8
    Sylvilagus (cottontail rabbits) 9–10
learning 51–53
    aggressive and nervous rabbits 34, 37–38
    classical conditioning 52
    clicker training 52
    instrumental/operant conditioning 52
    learned helplessness 53
    policy 52
    tonic immobility 53
    unpleasant experiences 52
Lepus 10–13
    L. alleni 12–13
    L. americanus 11
    L. arcticus 11
    L. californicus 13
    L. europaeus 11, 13
    L. nigricollis 13
    L. tibetanus 13
    L. timidus 10, 13
life outdoors 187–200
    adaptation, wild 187
    compost fencing 196
    Den Store Kaninboka 193
    exercise run 195
    garden area 187
    harness 198
    home-made and customized houses 193
    hutch and run systems 197
    isolated house 193
    mesh-fronted hutches 194
    mosquitoes 198
    optimal temperatures 187

life outdoors (*continued*)
    parasols and shades 194
    rabbits and grass 192
    rats and mice 198
    run 195
    safe area 192
    suitable and safe housing 192
    summer 190–192
        air conditioning 192
        checklist 192
        ears 191
        freezer blocks/bottles of water 192
        heatstroke symptoms 190
        high temperatures 190
        hindquarters 191
        humid and dirty substrates 191
        ventilation 190
    warrens 193
    winter 187–190
        bedding 188
        checklist 192
        climate changes 189
        cooler room 189
        floor 189
        fresh liquid water 188
        heating 188
        playing and grazing outside 188
        respiratory infections 188
        temperature conditions 187
        windbreak 188, 190
    wire-mesh flooring 194–195
litter tray 153–155
    customization 154
    disabled rabbits 154
    double-bottomed 154
    extended area 155–156
    fit 153
    hay 155
    Rabbit Litter Box 154
    urine drips 154

malocclusion 128–130
    cheek teeth 129–130
    and dental disorders 128–129
    development 86, 126
myxomatosis 92, 109–110
    biological control agent 4
    coin-sized raised scabs 109
    cutaneous 110
    distinctive disease 109
    full-blown 109–110
    mosquitoes 198
    pet population 109
    poxvirus 109
    and RVHD 5

    vaccination 109
    viral disease 109

nature and nurture 30–31
nervous rabbits 37
    'angry' rabbits 37
    cuddles 37
    freedom of movement 37
    physical conditions 34, 37
    plain language 37
    training sessions 37
    wild predecessors' behaviour 37
neurological disorders 105–107
    EC 105–107
    infected rabbit 106
    symptoms 105
    vestibular system 105
neutering 139–148
    adaptation, conditions 139
    and age-related questions 152–153
    Cheese and Biscuit 146
    definitions 140
    domesticated rabbits 139
    fat and lazy 148
    females 140–141
        endometrial hyperplasia 141
        endometrial venous aneurysm 141
        false pregnancies/pseudopregnancy 141
        hydrometra 141
        mammary lesions 141
        pyometra 141
        surgical procedure 141
        uterine torsion 141
        uterine tumours 140–141
    hormonal-related behaviour 143, 145
    life-saving supplement, starving families 139
    males 141–142
        chin gland lesions 142
        fertility 142
        testicular and prostatic lesions 141–142
    objections to/unnatural 139–140
    overpopulation 139
    rabbits' ovulation 145
    right time 142–144
    single rabbits 145–148
    social consequences 145
    social rabbits 59
non-steroidal anti-inflammatory drugs (NSAIDs) 111
nutrition 94–96, 117–137
    body condition, skinny rabbits 122
    bread 121–122
    changing diet 122–123
    checklist 95
    critical care 94
    dairy products 121
    dandelions 96

digestive diseases 123
digestive physiology 131–137
digestive system 117, 125
feeding strategy and behaviour 117
fluid therapy 96
food intake 95
GI 124
high-fibre pellets/nuggets 117
liquid pain relief drugs 96
liquidizing/blending 94
mashed pelleted diet 94
muesli and mixtures 121
and natural habitat 117
obesity 122
powder and warm water 94
pregnant and lactating does 122
'rabbit burrito' 95
rabbit's diet 117–121
and supply of liquid 94
syringe 94–95
teeth 125–131
temperature 96
toxic 122
water 123
young 122

Ochotonidae family (pikas) 7–8
origins and development 1–13
    Australia 4–5
    Belgian Hares 2
    companion rabbits 2–3
    environmental impacts 4
    European rabbit 5–7
    The House Rabbit Society 3
    hunting purposes 4
    Japan 5
    Lagomorphs 7–13
    medieval French monks 2
    myxomatosis 4
    Nordic rabbits 3
    pet 2–4
    Phoenicia 1
    The Rabbit Welfare Association & Fund 3
    Romans 1
    Spain and Portugal 5
    USA 5
    viral haemorrhagic disease 4
osteoporosis 102–105
    'angry' rabbits 103
    brittle bones 102
    cage dimensions 103
    calcification arthritis 103
    chronic pain 103
    cold and damp climate 103
    cramped conditions 103, 105
    exercise 103

    fracture and pain 103
    hiding pain 102–103
    hutch 104–105
    lameness 103
    living environment 103
    Oxbow Animal Health 103
    sedentary life 103
    splayleg 105
ovulation 202

peg teeth 7, 125, 128
pellets/nuggets 119, 120
    behaviour and feeding strategy 120
    bowl 120
    compressed and extruded 119
    fresh hay and water 120
    guinea pigs and chinchillas 120
    muesli versus pellets 120
    supplementary food 119
pet shops 26
pubertal rabbits 54–55

rabbit anatomy 83–87
    cardiovascular system 87
    dentition 86
    ears 85
    eyes 85–86
    fur 83–84
    gastrointestinal tract 87
    muscles and skeleton 84–85
    nose 86
    paws 84
    respiratory tract 86
    sexing and genitalia 87
    skin 83, 84
    tail 84
    urinary tract 87
rabbit calicivirus 4, 108
rabbit (viral) haemorrhagic disease (RVHD) 4, 5, 109
rabbit internal accomodation 168
rabbit selection 23–25
    baby/adult 23
    breeds 24–25
    male/female 23–24
    in pairs/single 24
The Rabbit Welfare Association & Fund (RWAF) 3, 171
rabbit-proofing of the home 177–185
    advantages 177
    babygate 178
    blankets 181
    change of mind 180
    corner protection 183
    Cottontail Cottage 183
    custom-made room 179
    delimited area 181–183

rabbit-proofing of the home (*continued*)
    distractions 181
    electricity challenge 179
    flexible protection 180
    food and beverages 183–184
    litter tray and water, cabinet 185
    location, power socket 180
    mouldings 181
    outdoors 178
    permanently mounted protection 180
    plants 181
    protective interior 180
    routines 178
    safe and suitable 179
    shoes and chargers 184
    spiral cables 180
    toilet seat 184
    toxins 181
    traditional farming 177
    vague signals 177
    walking, closed/closing doors 182–183
    wallpaper 181
    wire channels 180
    woodwork 181
rabbit's diet 117–121
    foods 117, 119–120
    fresh water 117
    grass and dried grass 117
    green leafy salad 117
    hay 117–119
    treats 121
    twigs and branches 121
    vegetables 120–121
rehoming and rescues 25–27
    breeders 25–26
    pet shops 26
    rabbit home 26–27
reproduction 201–203, 205
    abandoned and unwanted rabbits 201
    and breeding *see* breeding
    housing conditions and rabbit nest 202–203
    ovulation 202
    suitable rabbits 202
    time for relocation 205
    wild versus hutched rabbits 201
respiratory system 98, 110
    infection 98
    rabbit's life situation 98
    snuffles 98
    treatment 98
RVHD *see* rabbit (viral) haemorrhagic disease (RVHD)
RWAF *see* The Rabbit Welfare Association & Fund (RWAF)

skeletal and muscular disorders 102–105
snout to tail 83–114

abscesses 100–101
anaesthesia and nursing 110–113
    advantages and disadvantages 112
    blood pressure 112
    endotracheal tubes 112
    gas/injectable agents 112
    heating 112
    herbivore 110
    hypothermia 112
    hypoxia 110–111
    intravenous catheters 112
    nasal breathers 112
    NSAIDs 111
    nutritional support diet 112–113
    pain-relieving drugs 111
    pre-existing diseases 111
    premedication 111
    preparation and examination, hospitalized rabbit 110
    stress 110, 113
    supraglottic airway devices 112
    veterinarian 111–112
    visual examination 112
    warmth and fluids 112
cancer 107–108
diarrhoea 97
diseases 97–110
euthanasia 113
flystrike/fly larvae attacks (myiasis) 101–102
GT stasis 98
myxomatosis 109–110
neurological disorders 105–107
nutritional support 94–96
obesity 110
prey animals 83
rabbit 83–87, 92, 94
    anatomy 83–87
    healthy 92
    pharmacy and first aid 94
respiratory system 98
routine check-up 89–92
senior rabbits 92–93
skeletal and muscular disorders 102–105
skin diseases 100
sore hocks (pododermatitis) 102
subtle animals 83
symptoms of pain and illness 93
tonic immobility 113–114
travelling to clinic 88–89
urinary tract problems 98–100
veterinary clinics 88
VHD 108–109
social rabbits 20, 59–80
    and babies 75–76
    behaviour and biology 59
    bonding 59–64
    cohabitation 59

cohabitation and social need 72–74
    adult female 73
    advantages 72
    behavioural interactions 72
    behavioural needs 72–73
    leaping and jumping 74
    single-housed rabbits 73
    yawning, stretching and lying 74
combination 64–67
    other pairs 67
    siblings 64–66
    unfamiliar rabbits 64
commercial breeding systems 71
dogs and cats 78–80
environments and living conditions 59, 71
ferrets 80
and guinea pigs 76–78
housing in research 67
infectious disorders 71
neutering 59
neutral area 59
in practice 74
single rabbits in practice 74–75
stability 67–69
    bonded pair 69
    fighter's reunion 69
    grief 69
    The Private Life of the Rabbit 67
and wild rabbit 70–71
sore hocks (pododermatitis) 102, 103
sound 44
    cracking jaws 44
    grinding teeth 44
    grunt/growl 44
    hum 44
    rabbit scream 44
    teeth, quick clattering 44
supplementary food 119–120
    compressed and extruded pellets 119
    grass and nutritional composition 119
    muesli versus pellets 120
    pellets/nuggets 119, 120
Sylvilagus (cottontail rabbits) 9–10
    S. floridanus 9
    S. palustris 10

teeth 125–131
    abscesses 131
    cheek teeth 129–131

    dental disease *see* dental disease
    grinding 44
    quick clattering 44
    routine check-up 90–91

urinary tract problems 98–100
    anti-inflammatories 100
    bladder infections 99–100
    cystitis 98–99
    diet 100
    incontinence 98
    kidney/bladder stones (urolithiasis) 98, 99
    musculoskeletal 99
    neurological origins 99
    Oxbow Animal Health 100
    physical flushing of excessive sediment 100
    scalding 98, 99
    Trixie 100
uterine tumours 99, 107, 140–141

viral haemorrhagic disease (VHD) 92, 108–109
    biosecurity measures 108
    mosquito and flea control 109
    rabbit calicivirus 108
    RVHD 108, 109
    unvaccinated susceptible animals 108
    vaccination 108

wild rabbit 70–71
    colonies 44, 70
    domestication 70
    eating grass 117
    European 5, 9, 162
    frictions and quarrels 71
    gregarious and social animals 70
    habitat 161
    versus hutched 201
    hutches 70
    nutritious food first 32
    observation 71
    and predators 195
    proximity and distance, colony members 71
    social conduct 72
    underground tunnels and burrows 42
    VHD 108
    water 123